AF601439

UNITEXT for Physics

UNITEXT for Physics series publishes textbooks in physics and astronomy, characterized by a didactic style and comprehensiveness. The books are addressed to upper-undergraduate and graduate students, but also to scientists and researchers as important resources for their education, knowledge, and teaching.

Gleb Vdovin

Elementary Technical Optics

Springer

Gleb Vdovin
Delft, The Netherlands

ISSN 2198-7882 ISSN 2198-7890 (electronic)
UNITEXT for Physics
ISBN 978-3-032-08625-9 ISBN 978-3-032-08626-6 (eBook)
https://doi.org/10.1007/978-3-032-08626-6

This Springer imprint is published by the registered company Springer Nature Switzerland AG
The registered company address is: Gewerbestrasse 11, 6330 Cham, Switzerland

To my family

Preface

This course serves as an introduction to technical optics and optical instruments within a broader engineering curriculum. It is loosely based on single-semester lectures I gave to Optomechatronics students at TU Delft and TU Eindhoven.

I have observed that student's foundational knowledge of physical optics varies significantly. To address this, the course includes a quick introduction to the most essential principles of physical optics, focusing on limited set of concepts widely used in designing and implementing optical systems. Attempting to cover a broader range of optics proved impractical, as students outside an optics major typically lack the time, resources, and specific motivation. Therefore, this course takes a more focused approach, emphasizing basic understanding and practical application of core principles rather than comprehensive coverage.

A firm grasp of fundamentals is more critical than ever in modern engineering practice. While computer simulations and artificial intelligence are powerful tools for solving problems, they cannot replace the human engineer in synthesizing novel technical solutions. This creative process involves generating concepts, quickly evaluating them, and deciding whether to proceed. This inherently uncertain search relies on basic principles, intuition, and engineering estimates—areas where computers offer limited help. The ability to simplify a problem to its essence and develop a "back-of-the-envelope" estimate is a vital and learnable skill. To cultivate this, each chapter includes problem sets. Most problems come with solutions that employ simplifications and approximations, which is why many problems use the verb "estimate" rather than "find".

This course assumes a basic knowledge of physics and calculus. The material prioritizes clarity and practical application over formal rigor, focusing on the essential physical concepts needed for effective engineering estimation. A body of practical information is also included to make the course immediately useful for everyday optical and instrumentation engineering.

Delft, The Netherlands
August 2025

Gleb Vdovin

Acknowledgements This work would not have been possible without the good people who entered my life exactly when I needed their support the most. First of all, my parents, who gave me both motivation and freedom, are allowing me to follow my own path and take the knocks that are essential for personal development.

I would like to acknowledge my professors at the Department of Quantum Electronics, ITMO Technical University in Leningrad, who taught us a kind of no-nonsense engineering that had to work with simple means under harsh conditions. In particular, I wish to mention Sergei Volkov, my graduate supervisor, who quite literally saved me from going in a very wrong direction.

Sergei Chetkin supervised my work at the General Physics Institute of the Soviet Academy of Sciences. He was followed by **Simon Middlehoek** from TU Delft, who made it possible for me to pursue optical projects in the Department of Electrical Engineering—an unusual but invaluable arrangement. I also learned a great deal about optics, microtechnology, and control theory from Lina Sarro, Paddy French, Chris Dainty, and Michel Verhaegen.

I am also indebted to Vitali Bezzubik, Vadim Kiyko, Nikolai Belashenkov, Alex Naumov, Gerard Mourou, Margaret Murnane, and Henry Kapteyn, with whom I had the privilege to work and from whom I learned during different stages of my career.

Finally, I am grateful to my current colleagues Oleg Solovyev and Seva Patlan, who possess the rare gift of politely engaging in spontaneous technical conversations regardless of the level of insanity I bring to them—including my frequent inability to properly formulate questions or fully grasp their well-considered answers.

Competing Interests The author has no competing interests to declare that are relevant to the content of this manuscript.

Contents

Chapter 1
Faces of Optics

Abstract Technical optics employs distinct models to describe optical phenomena in instruments. • *Geometrical Optics* models light as rays. In a uniform, isotropic medium, these rays travel in straight lines. When encountering reflection, refraction, or a non-uniform medium, their paths bend to minimize propagation time. • *Wave Optics* describes light as a transverse electromagnetic wave. This framework is essential for understanding diffraction, which fundamentally limits the resolution of imaging systems. • *Quantum Optics* treats photons as entities with both wave-like and particle-like properties (wave-particle duality), where a photon's energy and momentum are determined by its wavelength. All three models are utilized in the design and analysis of optical systems, with each providing critical insight into different aspects of system performance.

1.1 Faces of Light

Historically, several models of light have been developed. Even today, light is described in different terms depending on the context. The appropriate choice of physical model depends on the nature of the problem:

- *Geometrical optics* assumes that energy propagates along light rays, obeying Snell's law of refraction at the interfaces of different media. In gradient-index optical materials, where the refractive index and absorption vary continuously, geometrical rays follow complex trajectories, minimizing the propagation time between the source and the receiver. Geometrical optics and ray tracing are widely applied in the design of complex optical systems, particularly to minimize geometrical aberrations.
- *Wave optics* describes light as a transverse electromagnetic wave, with electric and magnetic components oscillating in orthogonal planes. Light propagation is represented either by diffraction integrals of varying detail or by partial differential equations in vector or scalar form. Wave optics is essential for describing diffraction and interference in both coherent and incoherent optical systems, including imaging instruments, spectrometers, lasers, and waveguides.

G. Vdovin, *Elementary Technical Optics*, UNITEXT for Physics,
https://doi.org/10.1007/978-3-032-08626-6_1

- *Particle and quantum optics* describe light using the principles of statistics and quantum mechanics, including the uncertainty principle. Quantum optics is applied in the analysis of energy transfer by light, and in the description of light sources and detectors.

No experiment proves that light can be described exclusively as particles, waves, or rays. Rather, light exhibits properties of all three, depending on the context.

After working through this chapter, the student is expected to understand the applicability of each model to engineering problems in optics.

1.2 Geometrical Optics

In the 17th century, the predominant mechanical model of the universe was applied to light, giving rise to corpuscular theory, in which light particles propagate along rays. Within this framework, Newton presented the first fundamental treatment of reflection, refraction, and color [1].

In this approximation, the trajectory l of a ray propagating from point A to point B through a medium with a spatially varying refractive index n satisfies the condition

$$\int_A^B n\,dl = \min\,. \tag{1.1}$$

In a uniform isotropic medium with refractive index n, photons propagate with velocity c/n, where c is the speed of light in vacuum, and the ray trajectories are straight lines, as follows from (1.1). Refraction at an interface between different media is governed by Snell's law:

$$n \sin\varphi = n' \sin\varphi', \tag{1.2}$$

where n and n' are the refractive indices of the two media, and φ and φ' are the incidence and refraction angles, respectively. The angle of incidence is defined as the angle between the ray and the surface normal at the point of intersection.

If in Eq. 1.2 we set $\sin\varphi' = 1$, the angle of refraction is $\pi/2$ and the refracted ray propagates along the interface. The corresponding incidence angle, known as the critical angle for total internal reflection (TIR), is given by

$$\varphi = \arcsin\left(\frac{n'}{n}\right). \tag{1.3}$$

For any $\varphi > \arcsin(n'/n)$ $(n > n')$, light is completely reflected back into the first medium and no energy propagates across the interface. TIR is widely used in optical engineering. For example, many prisms rely on uncoated faces as TIR mirrors. TIR provides 100% reflectivity, in contrast to coated mirrors, which always exhibit some losses. However, TIR elements introduce phase shifts between polarization states, so care is required when they are used in polarization-sensitive applications.

Since $ndl/c = dt$, where dt is the propagation time, Eq. 1.1 also implies that light follows the path of least time.

If a source emits total power P [W] uniformly in all directions, the power density (irradiance) I [W/m^2] at distance L is distributed over a spherical surface of area $4\pi L^2$, and is therefore

$$I = \frac{P}{4\pi L^2}. \tag{1.4}$$

An ideal imaging system transforms rays so that all rays emitted from a single object point converge to a single point in the image. Within the approximation of geometrical optics, an absolutely sharp point image can be formed by a lens. Deviations from this ideal case, when rays from one object point fail to intersect at a single image point, are called geometrical (or ray) aberrations.

A simple experiment, however, challenges the ray model. If light were merely a bundle of infinitely thin, independent rays, one could isolate a single ray from a parallel beam using a screen with a small aperture. In reality, a small aperture produces a divergent cone of light. The smaller the aperture (pinhole), the stronger the divergence. A pinhole with diameter comparable to or smaller than the wavelength λ acts as a point source, radiating in all directions within a hemisphere. This behavior is explained only by wave optics.

1.3 Wave Optics

The wave theory of light explains the scattering observed with small apertures. Electromagnetic waves emitted by a point source—approximated by a pinhole of size comparable to λ—form a spherical wavefront in a uniform isotropic medium. Here, the term *wavefront* is introduced geometrically, as a surface of equal optical path length, without invoking the wave nature of light.

Point sources in uniform isotropic media produce spherical wavefronts. In imaging, such a wavefront is transformed by an optical system to converge at a point, forming an image. However, due to the oscillatory nature of light, focusing is fundamentally limited, blurring the image and restricting both resolution and information capacity of optical systems.

The first systematic treatment of light as an oscillatory process was given by Christiaan Huygens [7].

According to wave theory, light propagates as a transverse wave consisting of oscillating electric and magnetic fields. In an isotropic, non-absorbing medium, the electric field vector is orthogonal to the propagation direction. This differs from scalar waves such as sound, where the pressure field is scalar-valued. The oscillation direction defines polarization states. For a linearly polarized wave with maximum electric field E [V/m] in a medium of refractive index n, the irradiance I [W/m^2] is

$$I = \frac{E^2 n}{2Z_0} = \frac{E^2 n}{2\mu_0 c} = \frac{E^2 n\epsilon_0 c}{2}, \tag{1.5}$$

where $c \approx 3 \cdot 10^8$ [m/s] is the speed of light in vacuum, $Z_0 \approx 377\ \Omega$ is the free-space impedance, $\mu_0 = 4\pi \cdot 10^{-7}$ $[T/A]$ is the vacuum magnetic permeability, and $\epsilon_0 = 1/(\mu_0 c^2)$ [F/m] is the vacuum permittivity. The electric field amplitude E can be expressed in terms of irradiance I as

$$E = \sqrt{\frac{2IZ_0}{n}} = \sqrt{\frac{2I\mu_0 c}{n}} = \sqrt{\frac{2I}{c\epsilon_0 n}}. \tag{1.6}$$

In many practical situations, the optical field can be treated as a scalar wave. In this approximation, polarization is neglected, and only the amplitude A and phase φ are considered. Then monochromatic field at position $\mathbf{r}$ and time t is

$$U = A \exp \mathrm{i}(\omega t - k\mathbf{r} + \varphi(\mathbf{r})), \tag{1.7}$$

where A is the amplitude, $\omega = 2\pi\nu$ the angular frequency, $k = 2\pi/\lambda$ the wavenumber, and φ a slowly varying aberration phase. The speed of light in vacuum is

$$c = \lambda\nu. \tag{1.8}$$

If light propagates in a medium of refractive index n, the wavelength becomes $\lambda_n = \lambda/n$, while frequency remains constant, giving the light velocity in transparent medium

$$c_n = \lambda_n \nu = \frac{\lambda\nu}{n}. \tag{1.9}$$

The Doppler effect describes the frequency shift due to relative motion: $\nu = \nu_0(1 + v/c)$, where $v > 0$ if the receiver moves toward the source, negative otherwise.

A monochromatic coherent wave with wavelength λ is altered by obstacles, producing diffraction. Important cases in technical optics include:

- *Circular aperture*: A plane wave passing through an aperture of diameter D produces a divergent beam. The angular divergence (between the axis and first diffraction minimum) is

$$\varphi_0 = 1.22\frac{\lambda}{D}. \tag{1.10}$$

The far-field normalized intensity distribution ($Z \gg D^2/\lambda$) is

$$I(\varphi) = I_0 \left[\frac{2J_1(\rho)}{\rho} \right]^2, \quad \rho = \frac{\pi D \sin\varphi}{\lambda}, \tag{1.11}$$

where J_1 is the Bessel function of the first kind. In the focal plane of a lens of diameter D and focal length F, illuminated by a collimated beam of irradiance I_0,

$$I(r) = I_0 \frac{D^4}{F^2\lambda^2} \left[\frac{\pi J_1(\rho)}{2\rho} \right]^2, \quad \rho = \frac{\pi Dr}{F\lambda}. \tag{1.12}$$

This defines the aberration-free point spread function (PSF).

- *Focal intensity*: The maximum irradiance in focus of a lens with focal length F and diameter D is

$$I_F = I_0 \left(\frac{\pi D^2}{4F\lambda} \right)^2 \approx 0.62 I_0 \frac{D^2}{(F\#)^2\lambda^2}, \tag{1.13}$$

where $F\# = F/D$ is the F-number.

- *Image formation*: The image intensity distribution is given by the convolution of the ideal object O with the system PSF I:

$$\int_S O(\boldsymbol{\rho}) I(\mathbf{r} - \boldsymbol{\rho})\, d\boldsymbol{\rho}, \tag{1.14}$$

where $\mathbf{r}$ and $\boldsymbol{\rho}$ are coordinates in the image plane.

- *Slit diffraction*: A plane wave diffracted by a slit of width D produces divergence angle

$$\varphi = \frac{\lambda}{D}. \tag{1.15}$$

The intensity in the focal plane of a cylindrical lens is

$$I(x) \sim I_0 \left[\frac{\sin\rho}{\rho} \right]^2, \tag{1.16}$$

where $\rho = \pi Dx/(F\lambda)$ and x is the focal-plane coordinate.

- *Diffraction grating*: A polychromatic wave incident at angle ψ on a grating of period d produces diffracted orders at angles φ satisfying

$$d(\sin\psi + \sin\varphi) = k\lambda, \tag{1.17}$$

where $k = 0 \ldots \pm N$ is the diffraction order.

A coherent plane wave diffracted by an aperture of diameter D can be focused only within the near field. At larger distances, corresponding to the far field, focusing is no longer possible. The approximate boundary Z_{FF} is

$$Z_{FF} \gg \frac{D^2}{\lambda}. \tag{1.18}$$

In fact, a plane wave of irradiance I [W/m^2] diffracted by an aperture D reaches a maximum on-axis intensity of $4I$ at

$$Z_m = \frac{D^2}{4\lambda}.$$

Beyond Z_{FF}, irradiance decays approximately as Z^{-2}.

1.4 Quantum Optics

To make the description of light even more intricate, light can also be represented as a collection of photons. Each photon carries a discrete amount of energy, given by

$$E = h\nu = \frac{hc}{\lambda}, \tag{1.19}$$

where $h \approx 6.63 \times 10^{-34}$ [J·s] is Planck's constant. The momentum of a photon is expressed as

$$p = \frac{dE}{dc} = \frac{h}{\lambda}, \tag{1.20}$$

with the general laws of conservation applying both to photon energy and momentum. This particle description becomes essential when analyzing interactions between light and matter, as well as processes involving emission, absorption, and detection of light.

1.5 Fusion of Models

Geometrical optics, in which diffraction effects are neglected, is an indispensable tool for the design and composition of imaging systems. Thanks to its conceptual simplicity, it allows efficient analysis and design of complex assemblies composed of spherical and aspherical lenses and mirrors.

Wave optics, in contrast, is crucial for analyzing diffraction and interference, which determine the fundamental limits of optical resolution, the information capacity of an optical system, and the overall image quality. Applications include imaging instruments, laser resonators, spectral devices, optical fibers, and interferometers. These systems are typically described using a combination of geometrical and wave optics. In many cases, diffracted light may be treated as a geometrically divergent beam, with parameters defined by expressions (1.10) and (1.12). These expressions establish the necessary connection between geometrical and wave optics, a connection we will frequently use.

The quantum particle model becomes indispensable for describing light–matter interactions, the operation of optical sensors, the principles of spectral instruments, and for solving photometric and radiometric problems. It also plays a key role in the analysis of radiative heat transfer and even in futuristic concepts such as spacecraft propelled by light pressure.

Comprehensive treatments of the fundamental aspects of optical science—including geometrical and wave optics—are provided in [2]. Modern perspectives on optical science, incorporating quantum optics, are given in [3–5].

1.6 Conclusion

Geometrical optics describes light propagation in terms of rays, which follow paths of least time according to Fermat's principle. This model explains reflection, refraction, and total internal reflection, and remains the basis for the design of imaging systems. It also describes geometrical aberrations, which arise when rays from an object point fail to converge to a single image point.

Wave optics extends this description by treating light as a transverse electromagnetic wave. It explains diffraction and interference, which limit the resolution of optical instruments. Diffraction at apertures determines beam divergence, shapes the point spread function, and defines the resolving power of telescopes and microscopes.

Quantum optics introduces the photon, carrying discrete energy $E = h\nu$ and momentum $p = h/\lambda$. This model describes photon emission, absorption, and detection processes, as well as radiative pressure and radiation energy transfer. Applications include the operation of lasers, detectors, photometric devices.

Each model applies within defined limits. Ray tracing enables efficient design, wave optics sets the fundamental performance boundaries, and quantum optics govern interactions with matter. Combined, they provide the complete framework used in optical engineering.

1.7 Problems

1.1 Using the expressions for the momentum of a photon, estimate the mechanical force F that is developed by the light ray with power P, by normal reflection on the mirror surface with reflectivity R. Assuming solar illuminance of 1376 W/m^2 and Earth albedo (reflectivity) of $R \sim 0.5$, estimate the solar wind force acting on the planet Earth.

Solution (1.1). The momentum of a photon equals to

$$p = \frac{dE}{dc},$$

where

$$E = h\nu = \frac{hc}{\lambda}$$

is the photon energy, c is the speed of light and λ is the wavelength. So

$$p = \frac{h}{\lambda}.$$

The force

$$F = \frac{dp}{dt}N$$

equals the change of momentum per second for all N photons. For one photon the momentum change is $2p$ for reflection and p for absorption. Illuminance I produces

$$N = \frac{I}{h\nu} = \frac{I\lambda}{hc}$$

photons per square meter per second, developing the pressure of

$$P = (1 + R)pN = (1 + R)\frac{I\lambda}{hc}\frac{h}{\lambda} = (1 + R)\frac{I}{c}.$$

The total pressure equals

$$P_{tot} = (1 + R)\frac{I}{c}\pi\rho^2,$$

where $\rho \simeq 6.371 \cdot 10^6$ m is the radius of Earth. Finally, the total force $F \simeq 8.7 \cdot 10^8$ N. This solution gives a rough estimate, in the assumption that Earth absorbs/reflects as a disk.

1.2 The solar constant is $I = 1376$ W/m^2, the distance from the Sun to Earth is $L = 150000000$ km. Calculate the power P of the Sun.

Solution (1.2). Using Eq. 1.4 we state:

$$P = 4\pi \mathrm{L}^2 I \approx 3.9 \cdot 10^{26}\ \mathrm{W}$$

1.3 A spacecraft is driven by solar sail which has surface density of $\rho = 1$ [g/m^2], and reflectivity $R = 0.8$. Assume the solar sail spacecraft starts from the Earth orbit and goes away to open space. Calculate the maximum velocity V that can be achieved by solar sail. Assume the solar irradiance equals $I = 1363$ [W/m^2] at the distance from Earth to Sun of $L_0 = 1.5 \cdot 10^{11}$ m. Neglect the solar gravitation.

Solution (1.3). The solar pressure P at distance L is given by

$$P = \frac{I(1+R)}{c}\left(\frac{L_0}{L}\right)^2,$$

then the total energy density E [J/m^2], accumulated by the sail after travel from distance L_1 to the infinity is given by

$$E = \frac{I(1+R)}{c}\int_{L_1}^{\infty}\left(\frac{L_0}{L}\right)^2 dL = \frac{I(1+R)L_0^2}{cL_1}.$$

The density of kinetic energy of the sail can be expressed as

$$E = \frac{\rho V^2}{2}.$$

Equating above expressions, we obtain the formula for maximum speed

$$V = \sqrt{\frac{2I(1+R)L_0^2}{cL_1\rho}}.$$

For the listed parameters, the maximum achievable velocity is somewhat unimpressive, on the universal scale, just about 50 [km/s]. Starting the sail from the Mercury orbit with $L_1 = 4.7 \cdot 10^{10}$ m increases the maximum achievable velocity to 86 km/s. Even with this velocity, it will take the sail more than 15000 years to reach the nearest star, at the distance of 4.5 light years from Earth.

1.4 Derive the Eqs. 1.2 from 1.1.

Solution (1.4). In Fig. 1.1 the ray propagating from A to B experiences refraction on the interface between n and n'. The optical path length described by Eq. 1.1 takes the form:

$$n\sqrt{X^2 + (Y-h)^2)} + n'\sqrt{X'^2 + h^2} = min. \tag{1.21}$$

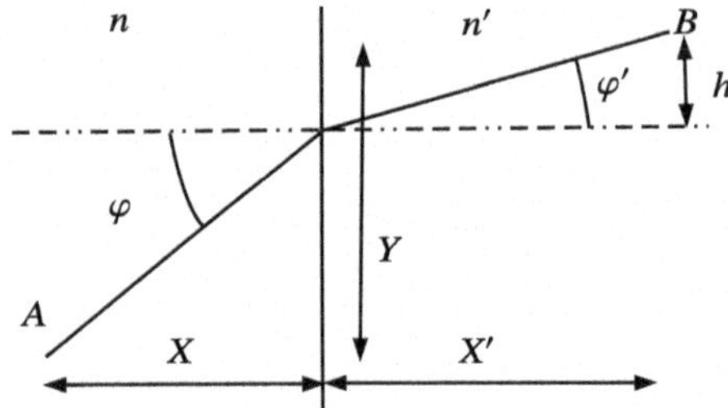

Fig. 1.1 To the derivation of Snells law

To achieve the minimum, the derivative of the optical path with respect to h should equal 0, from which directly follows:

$$\frac{h\,n'}{\sqrt{h^2+X'^2}} - \frac{(Y-h)\,n}{\sqrt{(Y-h)^2+X^2}} = n'\sin\varphi' - n\sin\varphi = 0 \tag{1.22}$$

1.5 In Sect. 1.3 we have stated:

- light rays are orthogonal to the wavefront,
- in a medium with refraction index n, light propagates with a velocity of c/n.

Derive the Snell's law using these two statements.

Solution (1.5). Consider propagation of the plane wave with wavefront AC from medium with refraction index n into medium with refraction index n', through the interface defined by straight line BC, as shown in Fig. 1.2. The speed of light in a transparent medium with refraction coefficient n is given by c/n, where c is the speed of light in vacuum.

As illustrated in Fig. 1.2, the time to cover the distance $|AB|$ in the first medium is equal to the time to cover the distance $|CD|$ in the second medium:

$$\frac{|AB|n}{c} = \frac{|CD|n'}{c}. \tag{1.23}$$

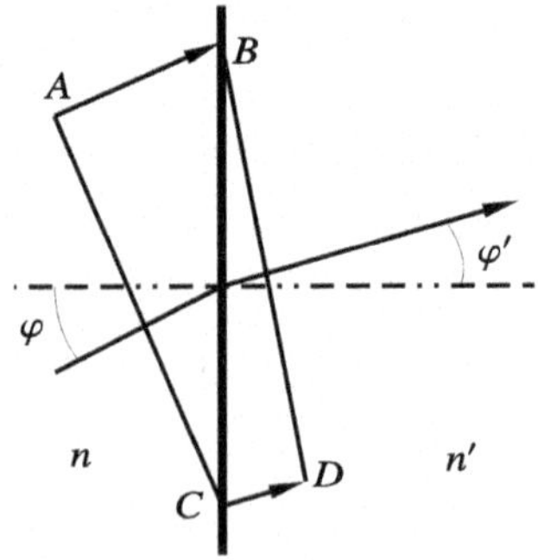

Fig. 1.2 To the derivation of the Snell's law

Also note that the phase delay at the distance $|AB|$ in medium with refraction index n is equal to the phase delay at the distance $|CD|$ in the medium with refraction index n':

$$\frac{2\pi|AB|n}{\lambda} = \frac{2\pi|CD|n'}{\lambda}. \tag{1.24}$$

Either of Eqs. (1.23) and (1.24) result in:

$$|AB|n = |CD|n'. \tag{1.25}$$

Dividing both sides of (1.25) by the common hypotenuse $|BC|$ we obtain:

$$n\frac{|AB|}{|BC|} = n'\frac{|CD|}{|BC|}$$

and accounting for

$$\frac{|AB|}{|BC|} = \sin\varphi, \text{ and } \frac{|CD|}{|BC|} = \sin\varphi',$$

we obtain

$$n\sin\varphi = n'\sin\varphi'. \tag{1.26}$$

The same expression can be derived from the Fermat principle, which states that "light travels between two points along the path that requires the least time, as compared to other nearby paths".

1.6 Calculate the wavelength of a lead shot with diameter of $D = 1$ mm fired with muzzle velocity of $V = 350$ m/s.

Solution (1.6). The energy E of the shot can be expressed as:

$$E = h\nu = \frac{hc}{\lambda} = \frac{mV^2}{2},$$

where ν is the frequency, λ is the wavelength, m is the shot mass. Substituting

$$m = \frac{\rho\pi D^3}{6},$$

where ρ is the lead density, we obtain

$$\lambda = \frac{12\,hc}{\rho\pi D^3 V^2}.$$

For $\rho = 11350$ [kg/m^3], we obtain $\lambda \approx 5.5 \cdot 10^{-25}$ m.

1.7 Stars have practically zero angular sizes, which is far below the resolution limit of the human eye. Nevertheless, we can clearly see them in the night sky, when the weather is clear. Explain how it can happen.

Solution (1.7). Due to diffraction on the eye pupil, the star image has angular size, limited by the eye resolution. The angular size of the star image is much larger then the star angular size. Nevertheless, if the star is bright enough, we clearly see the Airy disk, produced by the diffraction.

In practice people rarely see the classic airy disks, in most cases these images are distorted by the aberrations introduced by the atmospheric turbulence and the optical aberrations of the eye itself.

1.8 The angular size of Moon is $\varphi \approx 0.5^\circ$ and the distance from Earth to Moon is 384000 km. Calculate the number of points resolved on the Moon surface by a naked eye with pupil diameter of 6 mm, and a telescope with a diameter of $D_t = 1$ m. Assume $\lambda = 0.55\ \mu$m.

Estimate the diameter of ground telescope that would resolve the footprints of American astronauts.

Solution (1.8). Assuming $\lambda = 0.55\ \mu$m, the angular resolution of the eye is estimated as

$$\psi \approx 1.22\frac{\lambda}{D} = 2 \cdot 10^{-4}.$$

The angular diameter of the Moon, in radians, equals $\varphi \approx 0.5/57 = 9 \cdot 10^{-3}$. Then, the number of resolved points can be estimated as $(\varphi/\psi)^2 \sim 8000$. Please note that in practice the resolution of the human eye is limited by aberrations, so the above estimate is rather optimistic.

With a one-meter telescope, the angular resolution is estimated as

$$\psi_t \approx 1.22\frac{\lambda}{D_t} = 0.68 \cdot 10^{-6}.$$

Then, the number of resolved points can be estimated as $(\varphi/\psi_t)^2 \sim 1.75 \cdot 10^8$. However, such a high resolution is rarely achieved, as the air turbulence limits the resolution of the telescope.

For the astronaut's footprint, we can assume the footprint is 0.3 m long, then the angular size of the footprint is $\phi = 0.3/384000000 \approx 8 \cdot 10^{-10}$. Assuming $\phi = 1.22\lambda/D$, for $\lambda = 0.55\mu$m we find $D_t \approx 860$ m.

However, in all cases the calculated resolution will not be realized by a ground-based instrument due to the influence of atmospheric turbulence.

1.9 The faces of a plane-parallel plate have reflectivity r, the absorption in the plate is zero. Calculate the reflection R and transmission T coefficient of the plate.

Solution (1.9). The transmission for the initial ray equals $T_1 = (1-r)^2$. The transmission for the ray that has experienced single internal reflection equals $T_2 = T_1 r^2$. The transmission for the ray that has experienced n internal reflections equals $T_n = T_1 (r^2)^n$. The total transmission

$$T = \sum_{n=1}^{\infty} T_n = (1-r)^2 \sum_{n=0}^{\infty} (r^2)^n.$$

Using the formula for the sum of a geometric progression Appendix A.1:

$$\sum_{n=0}^{\infty} \alpha^n = \frac{1}{1-\alpha} \tag{1.27}$$

we obtain

$$T = \frac{(1-r)^2}{1-r^2} = \frac{1-r}{1+r}, \quad R = 1 - T = \frac{2r}{1+r} \tag{1.28}$$

This solution satisfies the energy conservation condition $T + R = 1$.

1.10 N plates described in Problem 1.9 are placed after each other. Calculate the reflection R_N and transmission T_N coefficient of the series.

Solution Consequently applying the solution (1.28) we obtain for two plates: $R_2 = 4r/(3r+1)$, for three plates: $R_3 = 6r/(5r+1)$. For N plates:

$$R_N = \frac{2Nr}{(2N-1)r+1}, \quad T_N = \frac{1-r}{(2N-1)r+1}. \tag{1.29}$$

The $2N$ multiplier comes because each plate has two mirror surfaces. Obviously, $R_\infty =$, and $T_\infty = 0$.

The expressions can be extended to the case of a stack of M single mirrors with reflectivity R and transmission $T = 1 - R$:

$$R_M = \frac{Mr}{(M-1)r+1}, \quad T_M = \frac{1-r}{(M-1)r+1}. \tag{1.30}$$

$T_M + R_M = 1$, and for an infinite number of reflectors $R_\infty = 1, \ T_\infty = 0$, in good agreement with the common sense.

1.11 N absorbing filters with absorption $a_1, \ a_2, \dots a_N$ are placed after each other. The light beam with power P_{in} enters the first filter, passes through all filters, and has power P_{out} after the last filter. Calculate the absorption coefficients $a_1, \ a_2, \dots a_N$ to obtain equal power absorption/dissipation in each filter.

Solution (1.11). Each filter absorbs and dissipates

$$P_f = \frac{P_{in} - P_{out}}{N}.$$

The beam power after passing through m filters equals to

$$P_m = P_{in} - mP_f,$$

so we can equate for the power dissipated in the m-th filter:

$$P_m a_m = P_f,$$

obtaining

$$a_m = \frac{P_{in} - P_{out}}{P_{in}N - (P_{in} - P_{out})m}$$

1.12 Lens with diameter D concentrates collimated light beam and produces maximum intensity I in focal spot. How the maximum focus intensity will change if we reduce the focal length by two times? How the maximum intensity I will change if we reduce the lens diameter D by 2 times?

Solution (1.12). The maximum focal intensity will be 4 times larger for a shorter focal length. The maximum focal intensity will be 16 times smaller for a lens with 2 times smaller diameter.

1.13 Assume the electric breakdown occurs in air when the electric field reaches the value of $E_b = 3 \cdot 10^6$ V/m. Coherent laser pulse with the wavelength of $\lambda = 1.06\,\mu$m and duration of $\tau = 1$ ms is focused through ideal lens with a diameter of $D = 1$ cm and focal length of $F = 0.1$ m. Assuming rectangular pulse shape and uniform power distribution in the aperture, estimate the maximum pulse energy P_{max} that would not cause the electric breakdown of air in the lens focus. Derive the general expression for the pulse energy to achieve the electric breakdown.

Solution (1.13). The power density (irradiance) I_a in the aperture:

$$I_a = \frac{4P_{max}}{\pi D^2 \tau},$$

Using Expr. (1.13), the irradiance in focus I_f is found as

$$I_f = \frac{\pi D^2 P_{max}}{4\,F^2 \tau \lambda^2},$$

with the electric field in focus, according to (1.6), given by

$$U = \sqrt{\frac{\pi D^2 P_{max} Z_0}{2 F^2 \lambda^2 n \tau}}.$$

Equating the electric field U to E_b and solving for P_{max} we obtain:

$$P_{max} = \frac{2 F^2 E_b^2 n \tau \lambda^2}{\pi D^2 Z_0} \approx 1.7\text{mJ}.$$

This result has mostly didactic value, as in practice the laser breakdown conditions have rather complex dependence on the laser pulse parameters [6]. The practical breakdown threshold can be different by orders of magnitude from the obtained result.

References

1. Sir Isaac Newton, *Opticks: Or, A Treatise of the Reflections, Refractions, Inflections and Colours of Light* (William Innys at the West-End of St. Paul's, 1730)
2. M. Born, E. Wolf, *Principles of Optics: Electromagnetic Theory of Propagation, Interference and Diffraction of Light*, 7th edn. (Cambridge University Press, Cambridge, 1999)
3. I.R. Kenyon. *The Light Fantastic: A Modern Introduction to Classical and Quantum Optics* (Oxford University Press, 2008)
4. G. Grynberg, A. Aspect, C. Fabre, C. Cohen-Tannoudji, *Introduction to Quantum Optics: From the Semi-classical Approach to Quantized Light* (Cambridge University Press, 2010)
5. E. Hecht, *Optics* (Pearson Education, Incorporated, 2017)
6. M.H. Niemz, Threshold dependence of laser-induced optical breakdown on pulse duration. Appl. Phys. Lett. **66**(10), 1181–1183 (1995)
7. C. Huygens, *Traité de la Lumière* (Pieter van der Aa, Leiden, 1690)

Chapter 2
Wavefronts, Rays, Imaging

Abstract This section introduces the fundamental components and parameters of an imaging system, including focal length, the system stop, entrance and exit pupils, and the diffraction-limited resolution. The pinhole camera is considered as an illustrative example, demonstrating that unlimited resolution is theoretically achievable with a simple pinhole, but the solution is impractical due to extreme size requirements and very low light-gathering ability of a pinhole. For lens-based systems, key parameters such as the field of view, numerical aperture, and F-number are defined in terms of system geometry. The paraxial (parabolic) approximation is introduced through a Taylor series expansion of a spherical wavefront. This approximation is then applied to derive the thin-lens equation. Parameters of an imaging system, such as transverse and longitudinal magnification are introduced. The section concludes with an estimate of the information capacity of an optical system as a function of its parameters.

2.1 Diffraction and Point Source

Consider a single point serving as the light source. In a uniform isotropic medium, geometric rays propagate in all directions at a velocity of c/n, following straight-line trajectories. Here, "uniform" means that the properties of the medium do not vary with spatial coordinates, while "isotropic" means that the properties are identical in all directions. By elementary geometry, there is only one straight ray passing through any point in space, connecting that point to the point source.

In this framework, a *wavefront* is defined as a continuous surface orthogonal to all rays emitted by the source. A point source in a uniform isotropic medium generates spherical wavefront.

The situation becomes more complex with two or more point sources. In this case, multiple rays pass through each point in space, making it impossible to define a single unique wavefront. However, rays can still be associated with their sources, so as many wavefronts exist as there are point sources. A collection of point sources defines an extended source, representing a real object. Conversely, any real object can

G. Vdovin, *Elementary Technical Optics*, UNITEXT for Physics,
https://doi.org/10.1007/978-3-032-08626-6_2

be decomposed into a set of point sources. This raises a practical question: how many point sources are required to reproduce the image of a given extended object? An infinite number of point sources is required in terms of geometrical optics. In reality, however, the number of resolvable sources is limited by diffraction and depends on the wavelength, the geometry of the optical system, and the angular size of the object. Due to diffraction, an optical system cannot resolve details smaller than the angular size

$$\varphi = 2.44\frac{\lambda}{D}, \tag{2.1}$$

where D is the aperture diameter of the optical system. Thus, $2.44\lambda/D$ may be taken as the effective angular size of a point source. Any source with angular size smaller than this can be considered a point source.

To summarize:

- The angular size of a point source is determined by the aperture diameter and the wavelength of light.
- No internal structure of the point source can be resolved.
- A point source produces a spherical wavefront across the aperture of observation.

If the field of view of the imaging system is S [steradian], the total number n of resolvable points may be estimated as

$$n \sim \frac{S}{\pi(1.22\lambda/D)^2}, \tag{2.2}$$

where the denominator approximates the solid-angle area of a single diffraction-limited image spot.

The optical field produced by an extended object is obtained by summing the contributions from all individual point sources, while accounting for diffraction and interference if the elementary sources within the object exhibit mutual coherence.

To image a single point, an optical system must transform the diverging spherical or plane wave into a converging spherical wave. A plane wave may be regarded as a limiting case of a spherical wave emitted by a point source at infinity. For engineering purposes, it is important to determine how "plane" a given wave actually is, and how far away "infinity" must be considered. A wave is effectively spherical within a given aperture if the peak-to-valley deviation from a reference sphere is much smaller than λ. Similarly, "infinity" is defined as a distance Z where diffraction effects dominate over geometrical propagation. The far-field distance Z_{FF}, serving as this effective boundary, is obtained from the condition

$$2.44\frac{\lambda}{D}Z_{FF} \gg D, \tag{2.3}$$

and, simplifying the inequality, may be expressed as

$$Z_{FF} > \frac{D^2}{\lambda}. \tag{2.4}$$

This boundary distinguishes the near-field (Fresnel) diffraction regime, described by the Fresnel transform, from the far-field (Fraunhofer) diffraction regime, described by the Fourier transform.

2.2 Pinhole Camera

Consider a point source S_1 located at infinite distance $L = \infty$ from a screen with aperture A, as shown in Fig. 2.1.

Since the source is at infinity, a bundle of parallel rays passes through the aperture and reaches the screen at distance L'. A light spot replicating the shape of the aperture appears in the image plane. Introducing a second source S_2 at infinity produces a second spot. These spots are *resolved* if the separation between their centers exceeds A, or equivalently if the angular separation between S_1 and S_2 satisfies

$$\varphi > \frac{A}{L'}.$$

The smaller the value of φ, the better the angular resolution of the system. By increasing L' or decreasing A, φ can be made arbitrarily small. In the purely geometrical approximation, therefore, the pinhole camera could in principle achieve unlimited resolution.

If the source distance L is finite, the ray bundle in the aperture diverges with angle A/L, and two sources are resolved if

$$\varphi > \frac{A}{L}.$$

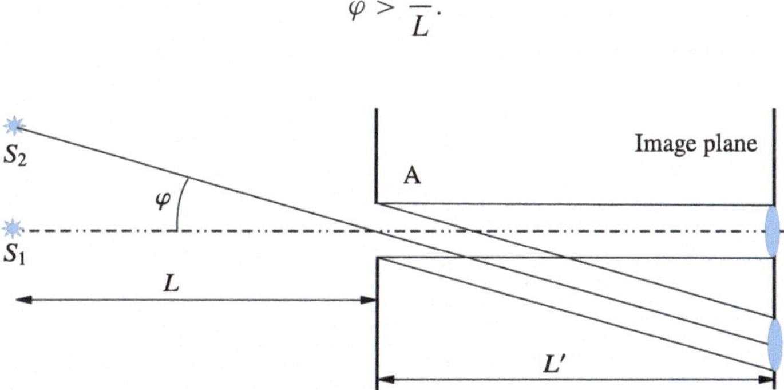

Fig. 2.1 Pinhole imaging in purely geometric approximation

Again, resolution can be increased arbitrarily by reducing A. However, two fundamental limitations arise:

- A small aperture transmits very little light, so the system becomes severely light-starved.
- Diffraction at the aperture produces a divergent light cone, with divergence angle $\varphi > 1.22\lambda/A$ for a circular aperture (Eq. 1.10), limiting the angular resolution to $\varphi > 2.44\lambda/A$.

To achieve high resolution with a pinhole camera, the aperture must be much larger than the wavelength. For two light spots to be separated on a screen at distance L' from the aperture, the condition must be satisfied:

$$\frac{1.22\,L'\lambda}{A} > \frac{A}{2}, \quad \text{or} \quad L' \geq \frac{A^2}{2.44\lambda}.$$

Thus, the parameters of a pinhole camera with angular resolution φ are given by

$$A = \frac{2.44\lambda}{\varphi}, \quad L' = \frac{2.44\lambda}{\varphi^2}, \quad F\# = \varphi^{-1}, \tag{2.5}$$

where A is the pinhole diameter, L' the distance from pinhole to sensor, and $F\#$ the F-number.

According to Eq. 2.5, unlimited resolution is theoretically achievable with a pinhole camera. However, practical considerations make such a system unrealistic: it would be enormous in size, would require extremely large image sensors, and would have poor light-collection ability due to a very large $F\#$. In a pinhole camera, light admitted through a small aperture is spread across a large sensor area.

These drawbacks arise because the pinhole camera relies on diverging ray bundles to form images. A more efficient approach uses an optical system that collects diverging rays from the source into a converging bundle at the image. To form a sharp point image, converging rays must intersect at a single focal point, corresponding to the center of curvature of the spherical wavefront. If the converging wavefront deviates from spherical form, rays fail to meet at a point, producing a blurred image. Hence, accurate imaging requires an optical system that converts diverging spherical wavefronts into converging spherical wavefronts with high precision, simultaneously for all object points.

Figure 2.2 shows an ideal imaging system. The space is divided into object space (left of the system) and image space (right). The point source emits isotropically, but the system aperture defines a cone of rays with angle φ, giving the *numerical aperture* (NA) $A = \sin\varphi$ in object space. After passing through the system, the converging cone has angle φ', with NA $A' = \sin\varphi'$ in image space. For an axially symmetric system, the straight line connecting an on-axis object point to its image is called the optical axis. Ideally, the optical system provides a one-to-one mapping: each resolved point in object space corresponds to a single point in image space. Geometrical optics

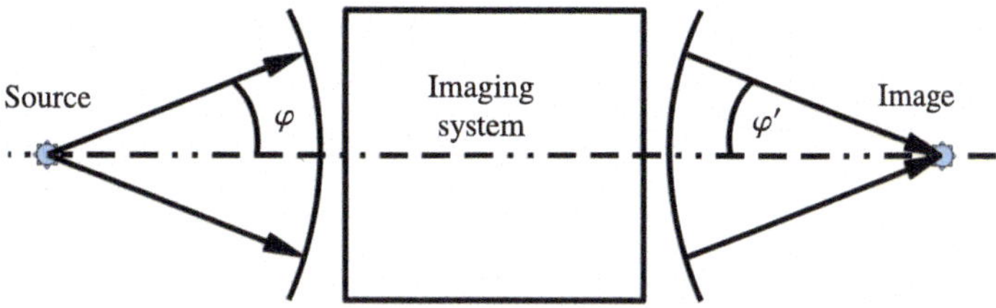

Fig. 2.2 An ideal imaging system converts a diverging spherical wave from a point source into a converging spherical wave forming the point-like image

establishes the fundamental laws governing this mapping and provides the design principles of optical systems.

If the object is extended, then the wavefront conversion should be performed for each resolved point of the object.

A spherical wavefront defines a homocentric cone of rays converging to a single point. Such homocentricity is a key requirement for aberration-free imaging.

2.3 Sign Conventions

Although the definition of signs for distances and angles in optics is conceptually straightforward, the existence of multiple conventions and approaches often makes the subject confusing. In this text we adopt the following rules (unless explicitly stated otherwise):

- Rays in an optical system are directed from left to right.
- Distances measured from left to right are positive. In some cases, distances are measured from a surface to a point; signs are reversed after reflection from a mirror.
- Object height is positive if the object lies above the optical axis, and negative otherwise.
- The radius of curvature of a surface or wavefront is positive if its center of curvature lies to the right of the surface, and negative if it lies to the left.

2.4 How Reflection and Refraction Influence the Wavefront

The imaging system introduced in Sect. 2 converts a diverging spherical wavefront into a converging one. Such transformations are achieved through the effects of reflection and refraction.

In geometrical optics, reflection is governed by the following rules:

- The reflection angle equals the incidence angle, both measured with respect to the surface normal at the point of incidence.
- At normal incidence, if the mirror surface is displaced by δ along the incident ray, the optical path length of the reflected ray changes by 2δ: light travels an extra δ toward the mirror and another δ after reflection. If a plane wavefront is described by $W_0 = \text{const}$ and the mirror surface by $f(y)$ (with y measured vertically in Fig. 2.2), the reflected wavefront is

$$W = W_0 - 2f(y), \tag{2.6}$$

where W_0 is the incident wavefront.

If a flat mirror is tilted by an angle ϕ, the reflected beam is deflected by 2ϕ. This "doubling" effect has important implications: the figure error of a mirror surface must be at least two times smaller than the allowable wavefront error.

For refraction, the situation is different. The velocity of light in a transparent medium (glass, plastic, fused silica, etc.) is c/n. When a ray passes through a plate of thickness d, the transit time is $\tau = nd/c$. During this time, light in free space would travel a distance $L = \tau c = nd$. Hence, the optical path difference (OPD) introduced by a refractive plate of thickness d is $(n - 1)d$. Thus, for a plane wave W_0 passing through a plate with variable thickness $d(y)$, the transmitted wavefront is

$$W(y) = W_0 - (n - 1)d(y). \tag{2.7}$$

Corrections introduced by reflective and refractive surfaces to the wavefront are additive. Constant terms, corresponding to the overall distance between source and optical system, are usually omitted since they can be absorbed into a shift of the origin.

Equations 2.6 and 2.7 highlight a key difference: surface errors of a mirror are doubled in the reflected wavefront, while errors of a refractive surface scale with $(n - 1)$. For typical optical glass ($n \approx 1.5$), this factor is about 0.5. Thus, to achieve the same wavefront precision, mirrors must be manufactured and aligned with roughly four times higher accuracy than refractive surfaces. These tighter tolerances increase technological complexity and cost, making precision mirror optics significantly more demanding.

2.5 Parabolic Approximation

Accurate description of the propagation and transformation of spherical wavefronts, particularly at large numerical apertures $A \gg 0$ in the object space or $A' \gg 0$ in the image space, requires nonlinear trigonometric expressions. However, a simpler linearized model can be introduced for rays and wavefronts propagating very close to the optical axis, at very small angles φ to the axis, with $A \sim 0$. In this regime, $\sin(\varphi)$ and $\tan(\varphi)$ may be approximated by φ itself.

In this approximation, a spherical surface of radius R, describing the wavefront W as a function of the transverse coordinate y, may be expanded into a Taylor series according to Appendices A.3 or A.2. Truncating the expansion after the quadratic term ensures that the parabola has the same curvature as the sphere at $y = 0$ (on the optical axis):

$$W = \sqrt{R^2 - y^2}\Big|_{|y| \ll R} = R - \frac{y^2}{2R} - \frac{y^4}{8R^3} + O(y^6). \tag{2.8}$$

In the parabolic approximation, only the first two terms of Eq. 2.8 are retained. The term in y^4 represents the first correction to the parabolic fit and corresponds to spherical aberration. Parabolic approximation is widely applied for practical engineering estimates:

- The sag Δ of a spherical surface of radius of curvature (ROC) R and semi-diameter a is estimated as

$$\Delta \approx \frac{a^2}{2R}, \tag{2.9}$$

valid for $a \ll R$.
- The angle φ of a ray belonging to a wavefront with radius of curvature R is approximated as a linear function of the transverse coordinate y:

$$\varphi = \frac{dW}{dy} \approx \frac{y}{R}. \tag{2.10}$$

- In the parabolic approximation, the phase ϕ [rad] of a spherical wavefront of curvature radius R is expressed as

$$\phi = n\frac{2\pi}{\lambda}\frac{y^2}{2R} = k\frac{y^2}{2R}, \tag{2.11}$$

where $k = 2\pi n/\lambda$ is the wave number in a medium of refractive index n.

The parabolic approximation is strictly valid only in an infinitesimally small neighborhood of the optical axis. Unlike a spherical wavefront, which converges homocentrically to a single point, a parabolic wavefront does not converge to a unique point; parabolic wavefronts are therefore not homocentric.

2.6 Lens Equation

If we place a point source at infinity and design the optical system to form its image at distance F in the image space, the system must transform an incoming spherical wave of infinite radius of curvature $R = \infty$ into a converging spherical wave of radius F.

In the parabolic approximation, and using Eq. 2.9, we express this as

$$\frac{y^2}{2\infty} = 0 \longrightarrow \text{Optical System} \longrightarrow \frac{y^2}{2F}. \tag{2.12}$$

Applying Eq. 1.7:

$$\exp \mathrm{i}\left(\frac{y^2}{2\infty}\right) \cdot \exp \mathrm{i}\left(\frac{y^2}{2X}\right) = \exp \mathrm{i}\left(\frac{y^2}{2F}\right),$$

or, in terms of wavefront contributions,

$$\frac{y^2}{2\infty} + \frac{y^2}{2X} = \frac{y^2}{2F},$$

from which we obtain $X = F$.

Thus, the optical system must introduce an additional wavefront modification of $y^2/(2F)$. The distance F in the image space, where the image of an infinitely remote object is formed, is called the *focal length.*

It is assumed that this optical path modulation occurs in a single *principal plane.* An optical system of this type is called a *thin lens*, and the intersection of the lens with the optical axis defines its *cardinal point.*

For a general case, an arbitrary input wavefront is transformed by adding the correction introduced by the optical system. For a source located at distance S, an optical system of focal length F produces an image at distance S' given by simple summation (Fig. 2.3):

$$\frac{y^2}{2S'} = -\frac{y^2}{2S} + \frac{y^2}{2F}.$$

From this relation we directly obtain the *thin lens equation*:

$$\frac{1}{S} + \frac{1}{S'} = \frac{1}{F}. \tag{2.13}$$

In Eq. 2.13, both distances S and S' are taken as positive, being measured from the surface to the corresponding point. If, however, we adopt the sign conventions

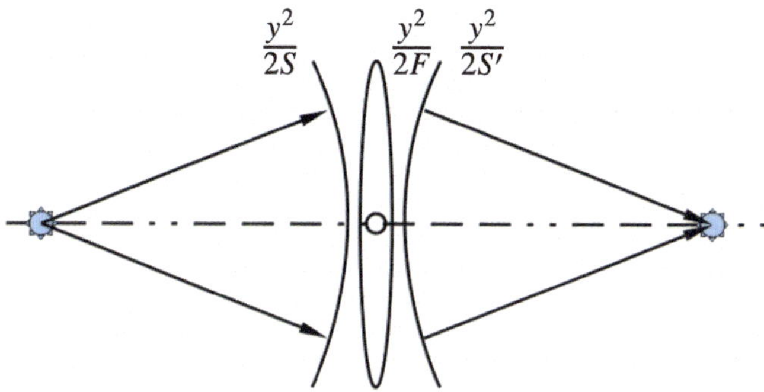

Fig. 2.3 A thin lens adds an OPD contribution of $\frac{y^2}{2F}$ to the divergent wavefront originating from a point at distance S in object space, described by $\frac{y^2}{2S}$, producing the convergent wavefront $\frac{y^2}{2S'}$ that focuses to a point at distance S' in image space

for radii of curvature from Sect. 2.3, we must take S negative and S' positive. The thin lens equation then takes the alternative form:

$$-\frac{1}{S}+\frac{1}{S'}=\frac{1}{F}. \tag{2.14}$$

Both forms are found in the literature, often without explicit clarification of the conventions used.

Now let us denote $S = a + F$ and $S' = b + F$, where a is the distance from the front focal point to the object, and b the distance from the back focal point to the image, as illustrated in Fig. 2.4. Substituting these into Eq. 2.13, we obtain

$$\frac{1}{a+F}+\frac{1}{b+F}=\frac{1}{F},$$

which reduces to another useful form of the thin lens equation:

$$ab = F^2. \tag{2.15}$$

This expression is valuable for calculating object and image positions relative to the focal planes.

The paraxial approximation and the thin lens model together provide a simple yet powerful framework that is routinely used to define the basic layout and estimate the engineering parameters of more complex imaging systems composed of multiple lenses and mirrors.

In practice, however, optical system layout requires tracing rays through real optical elements, with spherical or aspherical surfaces made from actual optical materials. The results may deviate significantly from the paraxial approximation, especially for objects located far off-axis or in systems with large numerical apertures.

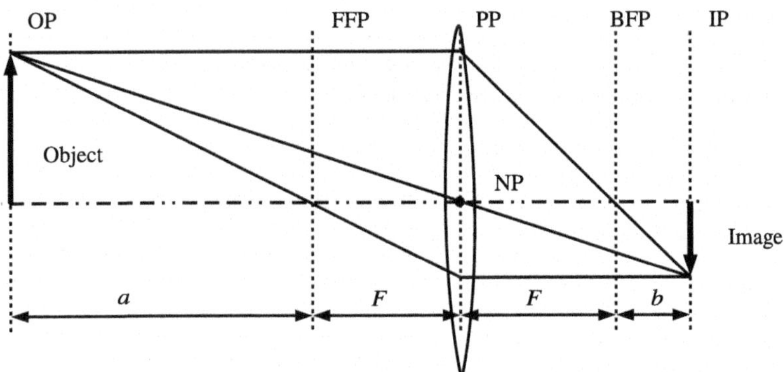

Fig. 2.4 A thin lens in air forms an image in the image plane (IP) of an object positioned in the object plane (OP). Rays passing through the nodal point (NP) do not change direction

2.7 Rays in Optical System

In axially symmetrical systems, the axis of symmetry coincides with the optical axis. Figure 2.2 illustrates such a system, with both the source and its image belonging to the optical axis.

Within this framework, we define several categories of rays:

- *Paraxial rays* propagate very close to the optical axis, at very small angles. Their propagation is described using the parabolic approximation.
- *Meridional rays* lie in the meridional plane, which contains the optical axis and the object point. If both the object and image points lie on the axis, all rays are meridional. These rays are relatively simple to trace.
- *Sagittal rays* lie in the plane perpendicular to the meridional plane.
- *Skew rays* do not lie in the meridional plane and do not intersect the optical axis. Tracing skew rays is considerably more complex than tracing meridional rays.

2.8 Geometry of a Thin Lens

A thin lens is an idealized optical component that converts a spherical wavefront of radius of curvature S into another spherical wavefront of radius S', according to the law (2.13):

$$\frac{1}{F} = \frac{1}{S} + \frac{1}{S'}, \tag{2.16}$$

where F is the focal length of the lens. For a thin lens, S and S' are the distances from the object to the lens and from the lens to the image, respectively.

The quantity

$$D = \frac{1}{F}$$

is called the focusing power of a lens (sometimes referred to as optical power, although this can be confusing since laser beams also have optical power expressed in watts). The focusing power is measured in diopters (d) [m^{-1}]. A lens with 1 d has $F = 1$ m, while a 10 d lens has $F = 0.1$ m.

In the special case of an object at infinity, $S = \infty$, the image distance is $S' = F$, and the image lies in the focal plane of the lens.

Figure 2.4 shows the principal points and planes of a rotationally symmetric lens. The axis of symmetry is the *optical axis*. Its intersection with the lens center is called the *nodal point* (NP) or principal point. Rays passing through the nodal point do not change direction (if the refraction coefficients in the image and object spaces are the same). The nodal point lies in the *principal plane* (PP), coincident with the lens. This plane separates space into the *object space* (to the left) and the *image space* (to the right). The *front focal plane* (FFP) and *back focal plane* (BFP) are located one focal length F from the principal plane. The intersections of the focal planes with the optical axis define the front and back focal points.

Paraxial ray tracing through a thin lens in air follows simple rules:

- Rays passing through the nodal point do not change direction.
- Any ray parallel to a nodal ray in object space is redirected at the principal plane so that it intersects the nodal ray in the focal plane. For a positive lens, rays intersect in image space, forming a real image. For a negative lens, their extensions intersect in object space, forming a virtual image.

From this geometrical treatment, several useful rules of thumb follow:

- A collimated beam[1] is focused to the focal plane. If the beam is tilted by angle β relative to the optical axis, the lateral position of its focal spot in the focal plane is $F\beta$.
- An object at infinity with angular size α produces an image of size $F\alpha$ in the focal plane.
- An object placed at distance $2F$ from the principal plane is imaged at distance $2F$ in the image space, with magnification $M = -1$ (inverted image).

In Fig. 2.4, the distances $S = a + F$ and $S' = b + F$, where a and b are the positive distances from the front focal plane to the object and from the back focal plane to the image, respectively.[2] The relation between a and b is

$$ab = F^2. \tag{2.17}$$

---

[1] A collimated beam has a plane wavefront; all rays are parallel in the geometrical approximation.

[2] In some sources, a and b denote distances from the nodal point to the object and image. Here we use S and S', since these values also correspond to the radii of curvature of the wavefronts at the principal plane.

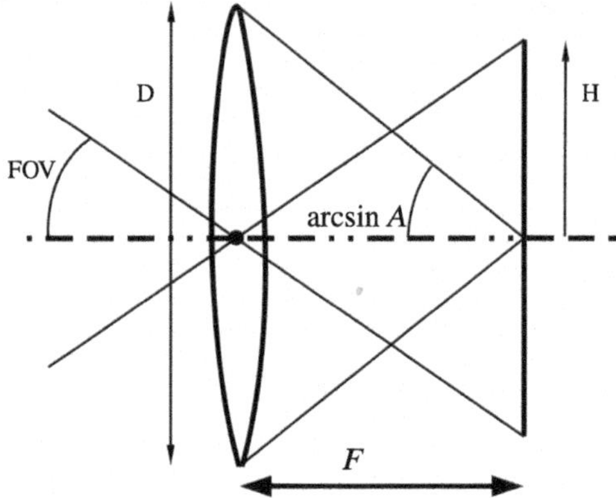

Fig. 2.5 Field of view (FOV) and numerical aperture A

Figure 2.5 illustrates two further important parameters:

- The *F-number* ($F\#$) of an optical system is the ratio of focal length to the diameter of the system pupil:

$$F\# = \frac{F}{D}. \tag{2.18}$$

- The *numerical aperture* (NA) A is the sine of the half-angle at which the lens aperture is seen from the axial image point. From geometry:

$$A = \frac{D}{\sqrt{4\,F^2 + D^2}}. \tag{2.19}$$

For small A, one may approximate $A \approx 1/(2F\#)$, but the exact relationship is:

$$F\# = \frac{\sqrt{1 - A^2}}{2A}, \quad A = \frac{1}{\sqrt{4F\#^2 + 1}}. \tag{2.20}$$

Obviously, in air $A \leq 1$ for all optical systems. For immersed microscope objectives, the refractive index n of the immersion medium is included in the definition of NA:

$$A_i = nA, \tag{2.21}$$

extending the range of possible numerical apertures to $0 < A_i < n$.

- The *field of view* (FOV) angle may be defined either as the ratio of object size to its distance from the nodal point, or as the ratio of image size to its distance from the nodal point.[3] For an object at infinity, FOV $= 2\arctan(H/F)$.

[3] In imaging optics, the FOV is often specified as the angle subtended by the diagonal of a full-frame sensor and the focal length: FOV $= 2\arctan(D/2F) = 2\arctan(H/F)$.

Finally, combining Eqs. 1.10 and 2.20, the diffraction-limited resolution in the image plane is

$$r \approx 0.61 \frac{\lambda}{A}. \tag{2.22}$$

2.9 Magnification

The lateral magnification M_t of a lens, defined as the ratio of image size to object size, follows from geometric similarity (see Fig. 2.4):

$$M_t = \frac{S'}{S} = -\frac{F+b}{F+a}, \quad \text{since} \quad a = \frac{F^2}{b}, \quad M_t = -\frac{b}{F} = -\frac{F}{a}. \tag{2.23}$$

The longitudinal magnification M_l is defined as the ratio of image length Δ to object length δ, both measured along the optical axis. Using (2.17) and (2.23), we obtain

$$\Delta = \frac{F^2}{a} - \frac{F^2}{a+\delta},$$

$$M_l = \frac{\delta}{\Delta} = \frac{a(a+\delta)}{F^2} \approx \left(\frac{a}{F}\right)^2 \Big|_{\delta\to 0} = \left(\frac{F}{b}\right)^2 \Big|_{\delta\to 0} = M_t^2. \tag{2.24}$$

Thus, in an ideal system the longitudinal magnification equals the square of the lateral magnification.

Equation 2.24 also shows that longitudinal magnification is a nonlinear function of the object depth δ. Strictly speaking, it is valid only for infinitesimally thin objects. For objects extended along the optical axis, the lateral magnification varies along their depth.

2.10 Stops and Pupils

With a non-zero FOV, the ray bundles from different points of the object travel through different volumes of space and through different parts of the optical system. Assuming all other surfaces do not limit the ray bundles, we can designate a single aperture to limit all ray bundles in the system. This aperture, frequently named system stop, has a property that all points of the aperture are common to all ray bundles passing through the system. Ray bundles, emitted by the separate points of an extended object, get together in the pupil, and then separate again as they propagate towards the image, as shown in Fig. 2.6. Thus, the pupil, shown in Fig. 2.6, is the only aperture in the system, that is common to all ray cones. The position of the pupil, defined by the system designer, has a strong influence on the paths of the

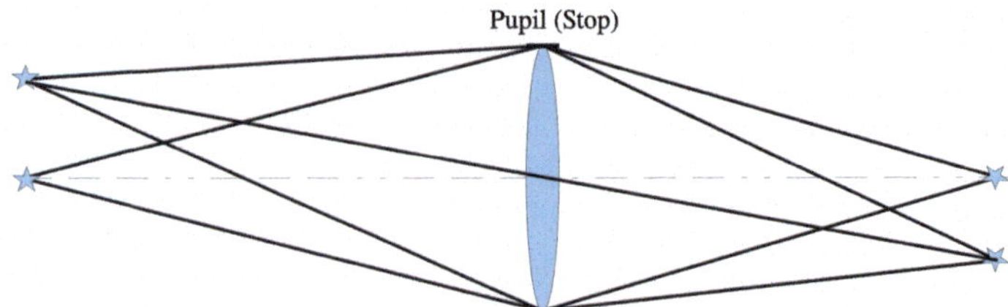

Fig. 2.6 The aperture in which the ray cones from all fields go through the same area is called system pupil

ray pencils and the general properties of the optical system. In practice, sometimes it is difficult to design and fabricate a system that limits the ray bundles only in one place, because other components of the system, before or after the stop, can also cut the ray bundles, especially for the points that are far away from the optical axis. This effect is called vignetting. In addition to darkening the corners of the image, vignetting changes the properties of the optical system, and has to be accounted for at the design stage (Fig. 2.10).

Since the optical system can include a number of optical components, intermediate images of the stop can be formed inside and/or outside of the optical system by these components. These pupil images are conjugated to each other and the ray bundles in the pupil images are copies of the ray bundle in the system pupil. This "copy" property is used in practical design. For instance, if the designer needs to send a ray into some position in the system pupil, which is positioned deeply inside the optical system, the ray can be sent to the conjugated point in the entrance in the entrance pupil, which is usually positioned in front of the system. Since all pupil images are conjugated to each other, the ray will be "copied" into the correct place in all the pupil images that are present in the system, and finally into the exit pupil.

The front part of the optical system positioned between the object and the stop forms the stop image that is called *the entrance pupil*. The back part of the system, positioned between the stop and the image, forms the image of the stop, called *the exit pupil*.

The entrance pupil is the smallest aperture of the optical system, visible from the on-axis point on the front (object) side. In a similar way, the exit pupil is the smallest aperture, visible from the on-axis point in the back (image) side of the system.

If the system stop is positioned in front of the system, the entrance pupil is co-incident with stop, as shown in Fig. 2.7A, and the exit pupil is represented by the stop image. If the system stop is positioned behind the system, the exit pupil is co-incident with the stop, as shown in Fig. 2.7B, and the entrance pupil is represented by the stop image. If the system stop is positioned inside the system, both the entrance and the exit pupils are formed by the images of the system stop, as shown in Fig. 2.7C. In general, optical system can have any number of pupil images, however it has only one entrance and one exit pupil.

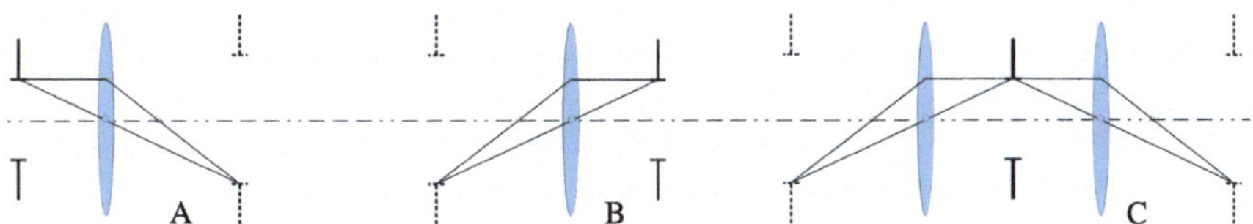

Fig. 2.7 Systems with different pupil configurations: the system pupil is coincident with the entrance pupil, the exit pupil is imaged (**A**); the system pupil is coincident with the exit pupil, the entrance pupil is imaged (**B**); the system pupil is inside the system, both the entrance and the exit pupils are imaged (**C**)

Many important properties of the optical system can be determined by tracing only two rays, representing the "boundary" cases of all possible ray bundles:

- The marginal ray with coordinates h, φ originates from the intersection of the object with the optical axis. It is aimed to the edge of the entrance pupil. Then it touches, consequently, the edges of all pupil images, including the exit pupil, and finally lands in the axial point of the image.
- The chief ray with coordinates $\widetilde{h}, \widetilde{\varphi}$ originates from the edge of the object. It is aimed to the nodal point, in the centre of the entrance pupil. Consequently it passes through all the images of nodal point in all intermediate pupil images, finally landing in the edge point of the image.

The position of the system stop has strong influence on the optical properties of the imaging system, as illustrated in Fig. 2.8.

In Fig. 2.8 the top system has the entrance pupil co-incident with the stop, and the exit pupil positioned at ∞, making all ray bundles exiting the system parallel to each other. Such a system is called image space (inverse) telecentric.

In the ordinary middle system the entrance and exit pupils are co-incident with the lens itself.

Finally, in the bottom system shown in Fig. 2.8 the entrance pupil is positioned at $-\infty$, so all ray bundles in the system entrance are parallel to each other, as they originate from this infinitely remote pupil, while the exit pupil is co-incident with the aperture diaphragm. Such a system is called telecentric.

Telecentric and inverted telecentric lenses are widely used in industrial machine vision and microscopy. In particular, telecentric systems are used in metrology, as it has the property to image object, positioned at different distances, in the same scale.

Figure 2.9 illustrates the special hypercentric system, in which the position of the aperture behind the back focal plane, makes possible observation of the object surfaces that are parallel to the optical axis. For example, it would image the sides of a cigarette, placed along the optical axis in front of the lens.

Since the entrance and exit pupils are not real material objects, but the images of the system stop, they can be positioned anywhere in the range $\pm\infty$.

As any other optical system, the human eye also has the entrance pupil. In observation instruments, working together with the eye, the size and the position of the exit pupil of the instrument should be matched to the size and the position of the entrance

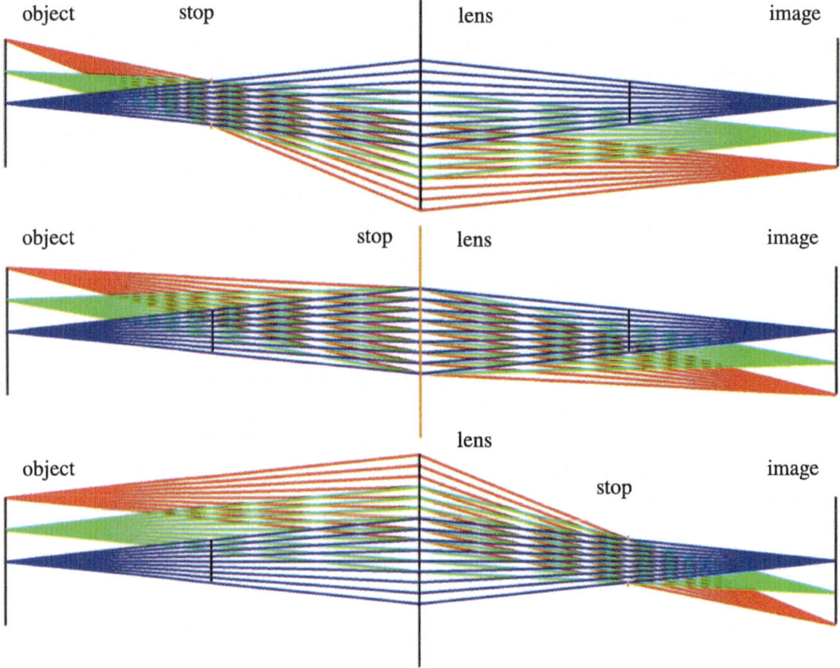

Fig. 2.8 Ray tracing through a single lens imaging system with different positions of the aperture diaphragm: image-space telecentric with stop in the focal plane in front of the lens (top), normal, with stop in the lens itself (middle), and object-space telecentric, with stop positioned in the back focal plane (bottom)

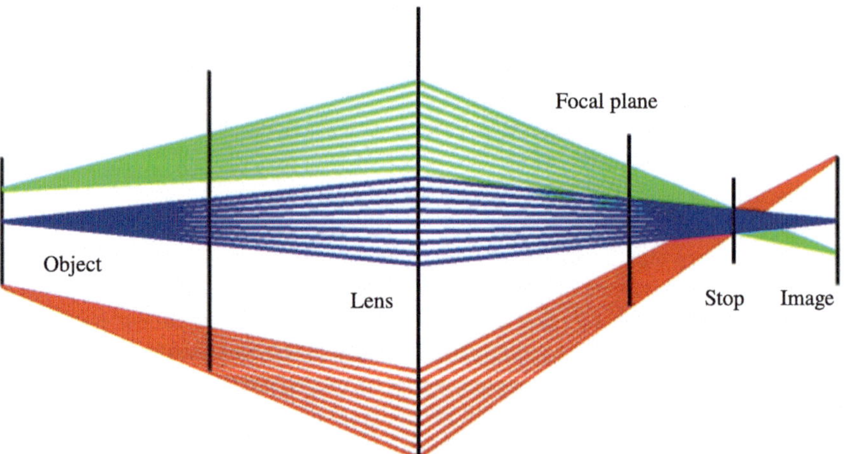

Fig. 2.9 Ray tracing through a hypercentric system

pupil of the human eye. For the comfort of the observer, binoculars, viewfinders, rifle sights have extended distance (also called eye relief) from the last component to the exit pupil, which should be coincident with the pupil of human observer for best visual comfort. The size of the exit pupil should be matched to the size of pupil of the eye. If the exit pupil of the instrument is much larger than the pupil of the eye, the eye effectively limits the system aperture, reducing the diffraction limited resolution of the instrument. If the eye pupil is larger than the exit pupil, then the full aperture of the eye is not utilized, also leading to the reduction of diffraction-limited resolution.

In spectral instruments the pupil, or its image, should be optically conjugated, or co-incident, to the dispersion element (prism, grating).

In an adaptive optical systems the system pupil should be conjugated to the wavefront corrector both in its position and the size.

The system stop is not the only aperture, that limits the propagation inside the optical system. Some ray pencils, propagating at an angle to the optical axis, could be limited by the lens apertures, hoods and baffles as shown in Fig. 2.10. There is no vignetting for the rays originating from the source A, however the rays emitted by source C are cut by the top part of the filter at the system entrance.

The linear vignetting coefficient V is defined as the ratio of the part of pupil M, that is "missing" due to vignetting as shown in Fig. 2.10, to the total stop diameter D:

$$V = \frac{M}{D}. \tag{2.25}$$

The vignetting reaches its maximum at the edges of the field of view. Apparently, no light can pass through the system for the points of the object, for which $V = 1$.

The area vignetting coefficient can be defined as the ratio of vignetted pupil area to the total area. Calculation of such a coefficient can be quite involving.

Vignetting reduces the working aperture for the off-axis object points, causing significant drop of the image illuminance along the edges. Also, since the working aperture is reduced for oblique rays, the diffraction-limited resolution also suffers,

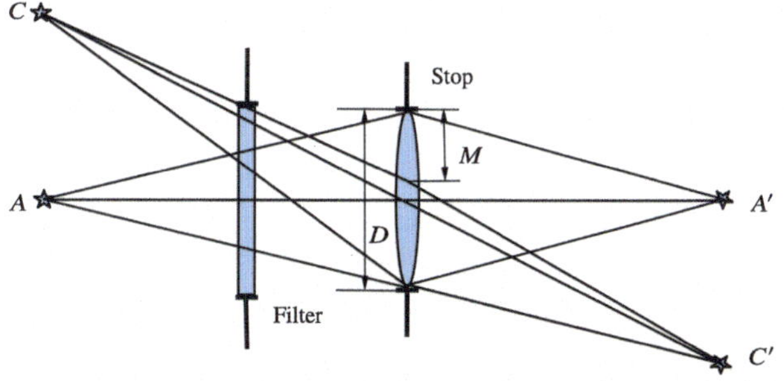

Fig. 2.10 Example of vignetting of off-axis rays by an aperture at the system entrance

resulting in some drop of image sharpness along the edges, especially in optical systems with perfectly corrected geometrical aberrations. However, in many optical systems the geometrical aberrations prevail. In such systems the vignetting can even play a positive role by effectively "stopping" the system for oblique rays, improving the edge resolution.

Vignetting is reduced when the size of the system stop is reduced. Imagine that the stop size in Fig. 2.10 is smaller than the position of the top ray that pass through the system from the source C. Such a small aperture will let through all ray pencils from both points A and C without any geometrical vignetting.

2.11 Depth of Focus

In the absence of aberrations, the diffraction-limited divergence $\varphi = 1.22\lambda/D$ defines the circle of confusion in the image plane of a lens, as illustrated in Fig. 2.11:

$$r_d \sim \frac{1.22\lambda S'}{n'D} = \frac{1.22\lambda F\#}{n'}. \tag{2.26}$$

In a purely geometric approximation, the focal image of a point source is an infinitely small spot. If the imaging plane is shifted by a distance Δ' from the focal plane, the radius of the geometric image becomes

$$r = \frac{\Delta'}{2\,F\#}, \tag{2.27}$$

due to divergence of the ray bundle. A point is considered positioned in acceptable focus if the geometric size r does not exceed the diffraction-limited size r_d. Equating (2.26) and (2.27) gives

$$\Delta' = \frac{2.44\,F\#^2\lambda}{n'}, \tag{2.28}$$

which represents the diffraction-limited depth of focus (DOF).

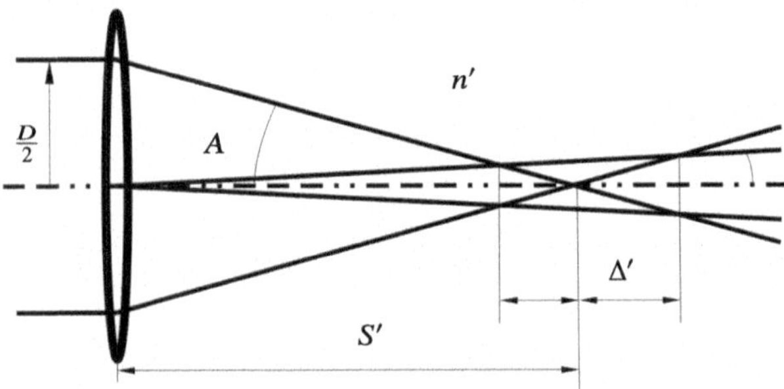

Fig. 2.11 Calculation of the depth of focus

To obtain the full depth of focus, we account for geometric widening on both sides of the focal plane:

$$2\Delta' \approx \frac{4.88\,F\#^2\lambda}{n'}. \tag{2.29}$$

Equation 2.29 is widely used in microscopy, as it defines the instrument's depth resolution and sectioning ability.

Microscope objectives designed for immersion in water or oil can have effective numerical apertures $A' \geq 1$. In this case A' equals the NA in air, multiplied by the refractive index of the immersion liquid: $A' = n'A$. Using $F\# \simeq 1/(2A)$ we obtain

$$2\Delta' = \frac{1.22\lambda n'}{(A')^2}. \tag{2.30}$$

In many systems the resolution is limited not by diffraction but by aberrations. Then the maximum acceptable circle of confusion r_d of Eq. 2.26 is replaced by a design parameter r_a, specified by technical requirements. For photographic lenses r_a is typically in the range of 5 to 50 μm. In this case, the aberration-limited depth of focus is

$$2\Delta' = 4r_a F\# \simeq \frac{2r_a}{A}. \tag{2.31}$$

The depth of field in the object space, DOF_{obj}, can be related to the image-space depth of focus $2\Delta'$. A rough approximation is obtained by dividing $2\Delta'$ by the longitudinal magnification defined by Eq. 2.24, but this method is inaccurate because longitudinal magnification is nonlinear. A more precise derivation is required.

2.12 Relation Between the Depth of Focus and the Depth of Field

As shown in Fig. 2.12, the image-space depth of focus Δ' corresponds to an object-space depth of field Δ. Applying Eq. 2.17 to the geometry of Fig. 2.12, the relation between Z and S' is

$$S' - F = b = \frac{F^2}{Z - F}, \tag{2.32}$$

where $ab = F^2$, $Z = a + F$, and $S' = b + F$. Similarly, the relation between $S' + \Delta'$ and $Z - \Delta$, corresponding to the far boundary of sharp focus, is

$$Z - \Delta - F = \frac{F^2}{S' + \Delta' - F}. \tag{2.33}$$

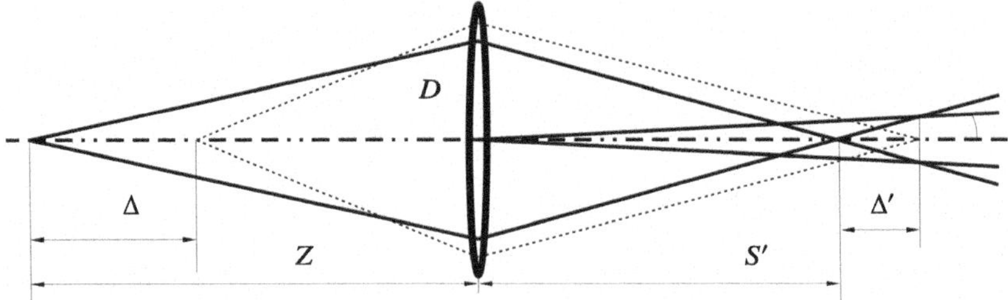

Fig. 2.12 Relation between the depth of field in object space and the depth of focus in image space

Substituting $\Delta' = \pm 2r_a(b + F)/D$ yields

$$Z - F - \Delta = Z - F + \frac{2\,Fr_aZ - 2r_aZ^2}{2r_aZ \pm DF}, \tag{2.34}$$

or equivalently,

$$\Delta = -\frac{2\,Fr_aZ - 2r_aZ^2}{2r_aZ \pm DF}. \tag{2.35}$$

If the system is focused at distance Z, Eq. 2.35 defines the boundaries of the sharp region in object space. The "+" sign corresponds to the near boundary, while the "–" sign corresponds to the far boundary.

When the system is focused at the hyperfocal distance Z_h, the depth of field extends from $Z_h/2$ to infinity. The hyperfocal distance is defined by setting the denominator of Eq. 2.35 to zero: $2r_aZ_h - DF = 0$. This gives

$$Z_h = \frac{DF}{2r_a} = \frac{F^2}{2\,F\#r_a}. \tag{2.36}$$

For diffraction-limited imaging, $r_a = 1.22\lambda/D$, yielding

$$Z_{dl} = \frac{D^2}{2.44\lambda}. \tag{2.37}$$

This diffraction-limited hyperfocal distance, Eq. 2.37, has similar physical meaning to the far-field boundary condition of Eq. 2.3.

For fixed r_a, the depth of field Δ depends strongly on focal length F. This explains why cellphone cameras with short focal lengths produce images with a large DOF, while cameras with long focal lengths yield much shallower DOF. In artistic photography and cinema, shallow DOF is often desirable to isolate the subject against a blurred background. In this case, lenses with long focal lengths and small $F\#$ are used.

Nowadays, similar effects can be simulated computationally. Modern smartphones employ two cameras separated laterally to capture slightly different viewpoints, producing parallax or depth-dependent displacement (as in stereoscopic photography). Depth information extracted from these image pairs is used to blur out-of-focus regions artificially through image processing.

2.13 Optical Prisms

The material of this section is based on [1, 2]. The majority of optical prisms are formed by flat glass surfaces to perform a predefined function in the optical system. These functions can be:

- Change the direction of the optical axis of the system to reduce the system dimensions, optimize the setup, or shift the entrance/exit pupil.
- Change the effective length of the optical path.
- Depending on the number of reflections, use the prism to introduce a mirror flip, reversal, or erection of the image.
- Merge and split optical beams or images, such as in binocular attachments in microscopy. A prism-based *camera lucida* was a popular drawing tool in the 19th century. The artist was looking through the prism simultaneously at the piece of paper and at the object. Precise sketches were then produced by outlining the contours of the object projected onto the paper.
- Multiplex and de-multiplex images. For example, split the image into three separate images projected onto three sensors sensitive to red, green, and blue, respectively.
- Perform retro-reflection, when the input beam is reflected exactly back regardless of its input angle.
- Rotate images.
- Enable angular scanning with wedge and drum scanner systems.
- Magnify the ray bundle in one or two directions to achieve an anamorphic effect.
- Disperse light into a spectrum.
- Polarize light.
- Introduce vignetting and aberrations into the ray bundle.

The number of reflections in a prism can vary from zero for a plane-parallel plate with thickness t, or an optical wedge with apex angle α, to N. Introduced into a converging beam bundle, the plane-parallel plate introduces a focus shift ΔS as shown in Fig. 2.13:

$$\Delta S = \frac{n-1}{n}t. \tag{2.38}$$

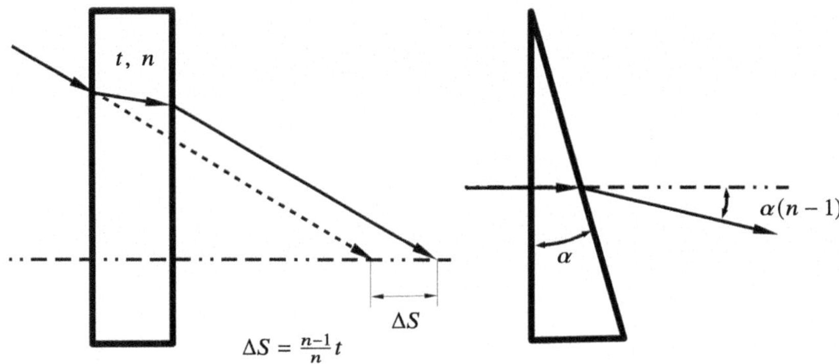

Fig. 2.13 Optical action of thin plate and wedge

The focus shift (2.38) should not be confused with the optical path difference $(n-1)t$, which is related to the phase delay $\Delta\varphi$ introduced by a thin plate:

$$\Delta\varphi = \frac{2\pi}{\lambda}(n-1)t. \tag{2.39}$$

An optical wedge with apex angle $\alpha \ll 1$ deviates the beam by $\Delta\alpha = (n-1)\alpha$ as shown in Fig. 2.13.

Figure 2.14 illustrates popular prism types. The Dove prism, with only one reflection, produces an inverted image and is used for image rotation. If the Dove prism is rotated around the optical axis by an angle α, the image is rotated by an angle of 2α.

The retro-reflecting prism with three reflective surfaces positioned at right angles to each other reflects light exactly back regardless of the prism tilt. Prisms with three surfaces are widely used to produce highly reflective targets and beacons. Prisms with two orthogonal surfaces (roof), similar to those shown in Fig. 2.14, are used as shutters for Q-switching of solid-state lasers. To achieve shutter action, the prism is rotated around the vertical axis. The laser feedback is switched when the prism takes the position at which it reflects light back into the resonator. The retro-reflecting property allows angular misalignment in the vertical plane to be avoided.

A double reflection, or any even number of reflections, produces a correct straight image.

The ray path through a prism can be analyzed by unfolding the prism into an equivalent glass plate or wedge. To produce the equivalent unfolded plate, the prism should be consecutively reflected with respect to each reflective face, as illustrated in Fig. 2.15 for the case of the Dove prism.

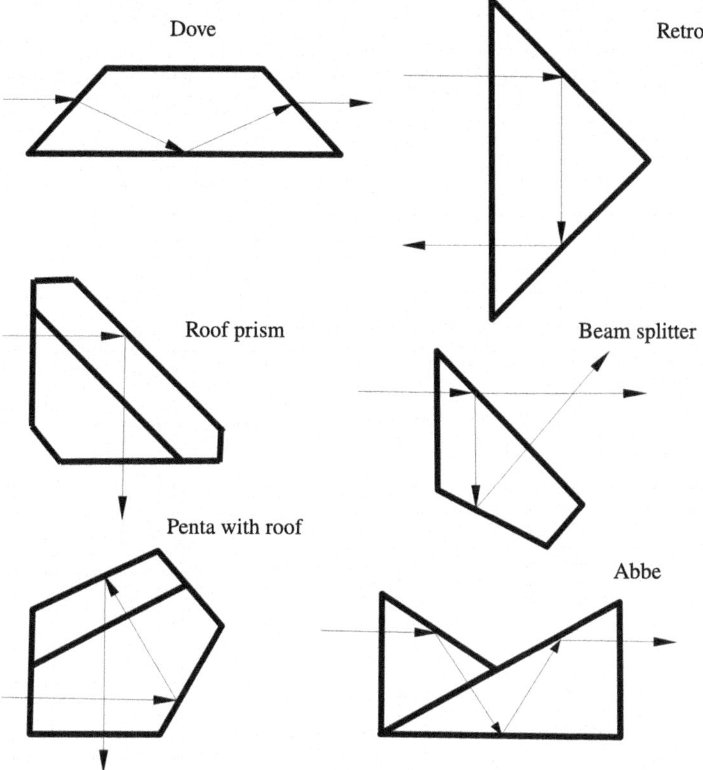

Fig. 2.14 Basic prisms

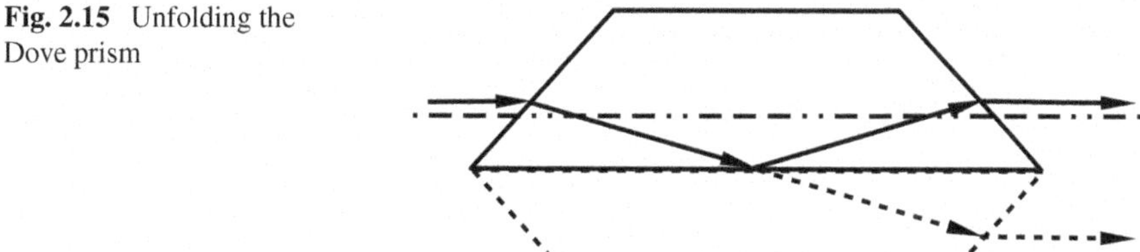

Fig. 2.15 Unfolding the Dove prism

2.14 Information Capacity of an Imaging System, Practical Considerations

Conceptually, the information capacity of an optical system is proportional to its ètendue and inversely proportional to the square of the wavelength. In practice, however, the information capacity is also limited by aberrations, light scattering within the system, and the signal-to-noise ratio (SNR) of the image sensor.

Consider an optical system with aperture D and focal length F, whose entrance pupil is illuminated by a plane wave with irradiance I [W/m^2] and wavelength λ. Assume that a square sensor with side length H' is positioned in the image plane. Further assume that part of the light entering the system is uniformly scattered by imperfections and parasitic reflections, producing on the sensor area $(H')^2$ an irradiance

$$I_s = k I_0 \frac{\pi D^2}{4(H')^2}, \tag{2.40}$$

where k is the fraction of scattered light. Anti-reflection coatings, baffles, and lens hoods are commonly used to reduce parasitic scattering.

The remaining, unscattered light is focused into the diffraction-limited spot, with intensity given by (1.13):

$$I_F = (1-k)\left[\frac{\pi}{4}\right]^2 I_0 \frac{D^2}{(F\#)^2\lambda^2}. \tag{2.41}$$

The initial SNR SNR_0 can then be estimated as the ratio I_F/I_s:

$$\mathrm{SNR}_0 = \left[\frac{\pi}{4}\right]^2 (1-k)\frac{4(H')^2}{\pi k (F\#)^2\lambda^2}. \tag{2.42}$$

However, Eq. 2.42 accounts only for the contribution from a single plane wave entering the aperture. In reality, every other wave entering the system also contributes to scattering, so the expression must be divided by the total number of resolvable input waves. This number is limited by the number of point sources the system can resolve, estimated as the ratio of S^2 to the area of the focal spot:

$$N \approx \frac{4(H')^2}{2.44\pi (F\#)^2\lambda^2}. \tag{2.43}$$

The effective SNR of the image is therefore

$$\mathrm{SNR} = \frac{\mathrm{SNR}_0}{N} \approx C\frac{1-k}{k}, \tag{2.44}$$

where C is a coefficient depending on the system design, the image spectrum, and the modulation transfer function (MTF).

In practice, additional non-optical noise sources–such as thermal noise, $1/f$ noise, and electronic noise–are always present. To account for them, Eq. 2.44 is modified as

$$\mathrm{SNR} = \frac{\mathrm{SNR}_0}{N} \approx C\frac{1-k}{k+\epsilon}, \tag{2.45}$$

where ϵ represents the contribution of other noise sources, normalized to the sensor's dynamic range.

The information capacity I of a diffraction-limited system is then proportional to

$$I = N \log_2(1 + \text{SNR}) \sim \frac{(H')^2}{(F\#)^2\lambda^2} \log_2\left(C\frac{1-k}{k+\epsilon} + 1\right). \tag{2.46}$$

Expression (2.46) shows that the information content depends on three principal factors: (1) the SNR, (2) the F-number $F\#$, and (3) the ratio H'/λ:

- The information transmitted by the optical system increases only logarithmically with SNR.
- For a given wavelength λ, the information capacity of a diffraction-limited system is proportional to the image area and inversely proportional to the square of the F-number, $F\#^2$.

In practice, the usable information is limited not only by scattering but also by other noise sources such as photon noise (at low intensities), amplifier noise, and quantization noise. Further information loss arises from optical aberrations, which reduce the number of resolved elements N compared to the ideal diffraction-limited case.

2.15 Conclusion

Imaging begins with the transformation of spherical or plane wavefronts emitted by object points into converging spherical wavefronts in image space. The resolution of such transformation is fundamentally limited by diffraction. A point source has an effective angular size of $1.22\lambda/D$, defining the minimum resolvable angular separation between object points and setting the limit for information capacity of the imaging system.

The pinhole camera illustrates the limiting case of geometrical imaging. In theory, arbitrarily high resolution is possible by increasing the pinhole size, but such a system would have enormous size and very bad light gathering capability.

Lens systems overcome these limitations by utilizing spherical wavefronts both in the object and image spaces. In the paraxial regime, spherical wavefronts are replaced by parabolic approximations, leading to the thin-lens equation $1/S + 1/S' = 1/F$, where S and S' are the object and image distances, and F is the focal length. This relation establishes the basis for imaging system layout. Optical magnification, both transverse and longitudinal, follows directly from this relation.

The system stop and its conjugate entrance and exit pupils govern the geometry of ray bundles, defining the numerical aperture, and the field of view. Placement of pupils and other apertures determines vignetting across the field. The diffraction-limited resolution in the image plane is defined $r \approx 0.61\lambda/A$, linking numerical aperture and F-number to system resolution.

Prisms act as assemblies of plane interfaces, performing beam deviation, reflection, inversion, retro-reflection, dispersion, polarization, and image rotation.

Finally, depth of focus and depth of field describe tolerances for sensor placement and object distance, with diffraction-limited expressions scaling as $\Delta' \propto \lambda(F\#)^2$. These parameters, together with aberration constraints and scattering losses, determine the usable information capacity of real systems.

2.16 Problems

2.1 Estimate the diameter of the optical system D, for which the Sun will look like a point source. The angular diameter ϕ of the Sun is approximately 0.5°, the observation wavelength is 550 nm.

Solution (2.1). The diameter D should satisfy to

$$1.22\frac{\lambda}{D} > 0.5\frac{2\pi}{360},$$

then $D < 77 \cdot 10^{-6}$ m.

2.2 Assuming the maximum wavefront error should not exceed $\delta = 50$ nm, estimate the limit $F\#$ and numerical aperture A at which parabolic approximation is still applicable to a wavefront with radius of curvature of $F = 3$ mm (corresponding to the focal length of a cellphone camera), $F = 50$ mm (corresponds to a classic 35 mm film camera), and $F = 57600$ mm (Hubble space telescope). Discuss the result.

Solution (2.2). Using the decomposition

$$\sqrt{F^2 - a^2} \simeq F - \frac{a^2}{2\,F} - \frac{a^4}{8\,F^3} - \frac{a^6}{16\,F^5} + \cdots$$

and limiting the solution to the third term only:

$$\frac{a^4}{8\,F^3} < 2\delta.$$

we obtain

$$a \simeq 2(F^3\delta)^{0.25}.$$

We used 2δ in the comparison because the wavefront deviates by δ to both sides of the nearest sphere. Then

$$F\# = \frac{F}{2a} \simeq 0.25\left[\frac{F}{\delta_{rms}}\right]^{1/4},$$

and, using (2.20) we obtain

$$A = \frac{1}{\sqrt{(2F\#)^2 + 1}} \simeq \left[0.25\sqrt{\frac{F}{\delta}} + 1\right]^{-0.5}$$

This solution is obtained for the amplitude error. For the residual mean square error δ_{rms}, the solution for $F\#_{rms}$ is slightly modified (feel free to derive):

$$F\#_{rms} \simeq 0.19\left[\frac{F}{\delta_{rms}}\right]^{1/4}.$$

If $\delta = 50 \cdot 10^{-9}$ we obtain $F\# \simeq 3.91$ for $F = 0.003$, $F\# \simeq 7.9$ for $F = 0.05$, and $F\# \simeq 46.1$ for $F = 57.6$ m.

2.3 Derive the paraxial formula to estimate the ray deviation φ introduced by a transparent glass wedge with refraction index n and apex angle $\alpha \ll 1$.

Solution (2.3). From Fig. 2.16 we state: $n\alpha \approx \varphi + \alpha$, then $\varphi = (n-1)\alpha$.

2.4 The wavefront of a light beam propagating in positive z direction, at $z = 0$ is described as

$$W = \frac{\rho^2}{2F} + \frac{\rho}{2F}, \tag{2.47}$$

where $\rho^2 = x^2 + y^2 < 1$, and $F = 10$. Schematically draw or describe the intensity distribution at the position $z = F$.

Solution (2.4). The wavefront is represented by a combination of parabolic term (lens), that focuses into a point, and conical term (usually produced by optical elements called axicons) that spread the distribution into a circle. The intensity will be represented by illuminated ring with radius of 0.5 and the ring width of $\sim 10\ \lambda$, as shown in Fig. 2.17. Combination of a lens and an axicon is used in laser technology for laser cutting of round holes.

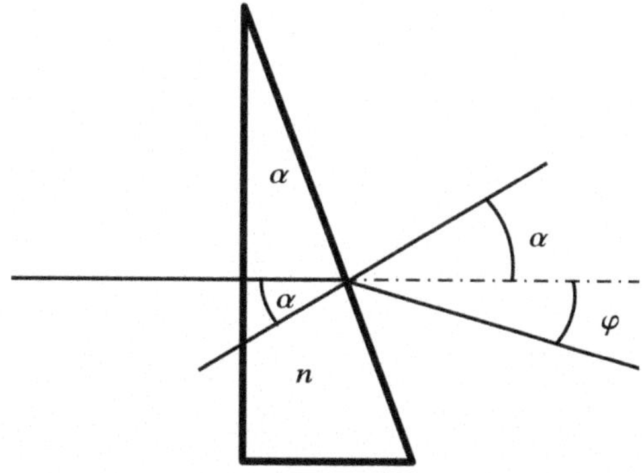

Fig. 2.16 Deviation of light ray by a wedge

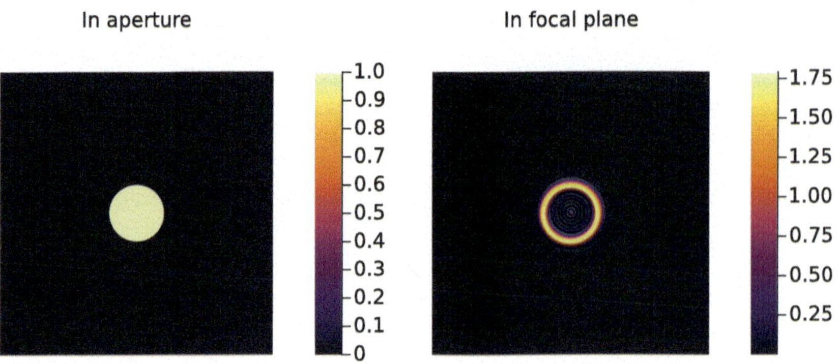

Fig. 2.17 Calculated intensity distributions in the plane of axicon (left), and in the focal plane (right)

2.5 Microsatellite is designed to make a photograph of moon surface in front of the probe before crashing vertically into the surface. It uses a camera with focal length of $F = 0.1$ m equipped with image sensor having 2000×2000 pixels with a pixel pitch of $P = 5\mu\text{m}$. The minimum camera exposure is $\tau = 1$ ms and the impact velocity is $V = 3000$ m/s. Assuming the image smear due to satellite movement does not exceed the value of pixel pitch P, calculate the distance Z to the surface, the maximum achievable system magnification and the time from the last sharp frame to the impact. The time to impact is important, as the probe has to transmit the information before being crashed.

Solution (2.5). To solve the problem, we have to compare two images. The first one is taken at a distance Z, the second one is taken at a closer distance $Z - \Delta Z$, where $\Delta Z = V\tau$. The pixel smear due to the difference between these two images must be smaller than the pixel size. The angular pixel size ψ is:

$$\psi = \frac{P}{F} = 5 \cdot 10^{-5}.$$

The maximum angular field φ equals

$$\varphi = \frac{A'}{2F} = \frac{A}{2Z} = 5 \cdot 10^{-2},$$

where A' is the sensor size, A is the field of view on the moon surface, and Z is the distance to the surface. Since the change of angular field due to change of distance Z should not exceed the angular pixel size, we can equate:

$$\frac{d\varphi}{dZ}\Delta Z < \psi,$$

or

$$\frac{A}{2Z^2}V\tau < \frac{P}{F}, \quad \text{or} \quad \frac{A'}{2FZ}V\tau < \frac{P}{F},$$

from which we find:

$$Z > \frac{\varphi}{\psi}\tau V.$$

Substituting parameters of the problem, we find $Z = 3000$ m, magnification $M = -F/Z \approx -3.3 \cdot 10^{-4}$, and the time to impact is $T = Z/V = 1$ s.

2.6 Calculate the shift of the focal spot by a plane parallel plate with refractive index n and thickness D introduced into a spherical wave with radius of curvature R.

Solution (2.6). Figure 2.18 illustrates the ray passing through the plate. The incidence angle can be defined as $\alpha = h/R$, where h is the height of the ray. When the ray is passing through the plate, the exit coordinate will be shifted by $\delta = (\alpha - \varphi)D$. Then, the intersection of the ray with the optical axis is shifted by $\Delta = \delta/\varphi$. Finally,

$$\Delta = \frac{n-1}{n}D.$$

Since the shift does not depend on the incidence angle (only in paraxial approximation!), it will be the same for all rays converging to the focal spot.

If the rays are focused into point A, as shown in Fig. 2.18, and the plate is inserted between the focusing system and the focal plane, the focal plane will be shifted by Δ to the point B, further from the system.

2.7 Camera is focused to a star and produces sharp image. When the camera lens is shifted by $\Delta = 1$ mm from the image plane, the confusion circle has diameter of $d = 0.5$ mm. Estimate the $F\#$ of the lens.

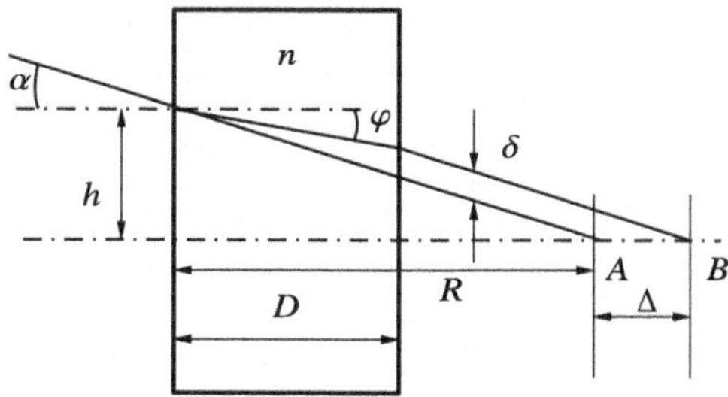

Fig. 2.18 Shift of focus, introduced by a plane-parallel plate

Solution (2.7). Assuming the focal length F is much larger than the lens shift, $F \gg \Delta$, we obtain

$$F\# \approx \frac{\Delta}{d} = 2.$$

If the shift is comparable to the focal length, then a more precise expression should be used:

$$\frac{F+\Delta}{D} = \frac{\Delta}{d},$$

where D is the lens aperture. Then

$$F\# = \frac{\Delta}{d}\frac{F}{(F+\Delta)},$$

and the result is dependent on the focal length of the lens.

2.8 James Bond has a fixed lens camera that is focused to the infinity. Bond needs to make a photo of a secret code book from a 0.5 m distance, to get it sharp in the right scale. His girlfriend has two pairs of spectacles: one has optical power of –1 diopter, the other one has optical power of +3 diopters. Bond suspects that he can use these glasses to make a sharp photo, but does not know how. Give him your advice.

Solution (2.8). Placing both spectacles in front of the camera lens results in an optical system focused to a distance of ~0.5 m.

2.9 Assuming the diameter of confusion circle is 20 μm, estimate the tolerance for the register distance for Leica film camera, for minimum F number of $F\# = 1.4$. The register distance is the distance between the objective flange and the sensor plane.

Solution (2.9). The tolerance should not exceed the depth of focus.

2.10 Camera has an autofocus mechanism covering object range from infinity to 1 m. This mechanism can be set to N different positions. Calculate the minimum number of auto-focus positions N for a smartphone camera lens with focal length of $F = 4$ mm and $F\# = 2$ assuming diffraction limited image quality. Do the same for a film camera lens with F = 28 mm and $F\# = 2.8$ assuming confusion circle diameter of 30μm. Assume $\lambda = 550$ nm.

Solution (2.10). For the cellphone lens the total travel to focus from the infinity to ~1 m equals

$$Z' = \frac{F^2}{Z - F},$$

where $Z = 1$ m and $F = 0.004$ m. The diffraction-limited DOF is given by (2.29)

$$2\Delta = 4.88\ F\#^2\lambda.$$

The number of steps is given by the integer ceiling of $Z'/(2\Delta)$.

$$N = \frac{\lambda F^2}{4.88(Z - F)F\#^2} = 1.49.$$

Autofocus with only two positions is sufficient for the cellphone camera.

The DOF of the film camera with $r_a = 15$ μm is given by

$$2\Delta = \frac{2r_a}{A} = 2r_a\sqrt{4\,F\#^2 + 1}.$$

Then, the number of steps is given by

$$N = \frac{F^2}{2r_a(Z - F)\sqrt{4\,F\#^2 + 1}} \approx 9.7.$$

The film camera should have at least 10 autofocus positions.

2.11 For an imaging system with a rectangular image size of $w \times h$ and numerical aperture $A \approx 1/2\text{F\#}$, using the expressions (1.10) and (2.22), derive the approximate formula for the number of resolved pixels, at wavelength λ. Assume the focal length F is equal to the image frame diagonal (normal lens). Estimate these numbers for $A = 0.2$, $\lambda = 550$ nm, and (1) the frame size of 2×3 mm, for a CCD camera, (2) for the frame size of 24×36 mm, which is the frame size of the standard film camera, and (3) for the frame size of 12×18 cm, which is the popular size of the photographic plate used in box cameras in the 19 and 20th centuries.

Assuming the scattering coefficient $k = 0.05$, estimate the information contents of the images produced by these cameras, using formula Eq. 2.46.

2.12 Compare the hyperfocal distance of two diffraction-limited lenses with numerical aperture of $A = 0.2$: one with $F = 4$ mm, the other one with $F = 40$ mm. Assuming square field with a diagonal of 45°, compare the information capacity of these lenses at $\lambda = 550$ nm, making reasonable assumptions for unknown parameters.

Solution (2.12). The lens diameter is given by

$$D = \frac{F}{F\#} = \frac{2\,FA}{\sqrt{1 - A^2}}.$$

The hyperfocal distance Z_h, for a diffraction-limited imaging, is given by (2.37):

$$Z_h = \frac{D^2}{2.44\lambda} = \frac{4\,F^2\,A^2}{2.44\lambda(1 - A^2)}.$$

Substituting the data we obtain for $F = 4$ mm $Z_h \approx 1.98$ m, and for $F = 4$ cm $Z_h \approx 198$ m.

Since the numerical apertures of both lenses are equal, both lenses provide the same linear resolution in the image plane. However, the image area of the lens with $F = 4$ cm is 100 times larger. So the information capacity of the lens with $F = 4$ cm is $\sim$100 times larger than that of a lens with $F = 4$ mm.

2.13 Trace rays and build the image of an object, placed in the object space halfway between the nodal point and the focus of a positive lens. Do the same for a negative lens. Explain the result.

2.14 Optical system with focal length $F = 5$ cm is focused to an object at a distance of $L = 2$ m in front of the system. Light filter, made of BK7 glass with $n = 1.515$, with thickness $D = 3$ mm is inserted. Calculate the shift of focal plane Δ if the filter is inserted

- behind the lens,
- in front of the lens.

Solution (2.14). Let a is the distance from the front focal plane to the object, and b is the distance from the back focal plane to the image. For the filter placed behind the lens the image plane is shifted by $\Delta b = D(n-1)/n \approx 1.02$ mm further from the lens

The filter inserted in front of the lens causes the shift of the visible position of the object by $\Delta a = -D(n-1)/n \approx 1.02$ mm, closer to the lens. Using the expression $ab = F^2$, we find

$$b = \frac{F^2}{a}, \text{ then } \Delta b \approx \frac{db}{da}\Delta a = -\frac{F^2}{a^2}\Delta a.$$

Substituting $a = 1.95$ m, $\Delta a = -1.02$ mm, we obtain $\Delta b = 0.65$ µm. The image plane is shifted less than by 1 micron from the lens. In a situation when the distance to object much larger than the focal length, the filter inserted in front of the lens in the object space does not change much in the optical system, while the filter inserted behind the lens, in the image space, requires the system to be re-focused.

2.15 Thermal deformation has caused the focal spot formed by the lens with focal length F to shift from the point $z = F,\ y = 0$ to the point $z = F + \Delta,\ y = \delta$. The optical axis of the system is coincident with the z axis, direction of y is orthogonal to the optical axis. Find the corresponding expression for the wavefront aberration $w(x, y)$, using parabolic approximation.

Solution (2.15). In the paraxial approximation the wavefront is described by

$$W(x, y) = \frac{x^2 + y^2}{2F}.$$

The shift by Δ along the z axis corresponds to the wavefront deformation of:

$$\frac{dW}{dF}\Delta = -\frac{x^2 + y^2}{2F^2}\Delta,$$

In a similar way, the shift by $delta$ along the y axis corresponds to the wavefront deformation of

$$\frac{dW}{dy}\delta = \frac{\delta}{F}y.$$

Finally, the aberration is described by:

$$w(x, y) = \frac{\delta}{F}y - \frac{x^2 + y^2}{2F^2}\Delta,$$

2.16 A thin lens with focal length F produces a sharp image of the object. The distance between the object and the image is L. The lens is shifted along its optical axis by a distance l, and again, it produces sharp image of the same object in the same plane. Calculate F,

Solution (2.16). Let a and b be the distances from object to lens and from the lens to image. Apparently $a + b = L$, $a - b = l$, and $1/a + 1/b = 1/F$. Solving this system we obtain $F = (L^2 - l^2)/(4L)$.

2.17 Parabolic mirror with focal length F is made of material with thermal expansion α. Calculate the change of focus ΔF, caused by uniform temperature change ΔT.

Solution (2.17). The focal length is given by

$$F = \frac{a^2}{2\delta},$$

where a is the mirror radius and δ is the mirror sag. After the temperauture has changed, the new radius is given by $a(1 + \alpha\Delta T)$, and the sag is $\delta(1 + \alpha\Delta T)$. Then

$$\Delta F = \frac{(a \cdot (1 + \alpha\Delta T))^2}{2\delta \cdot (1 + \alpha\Delta T)} - \frac{a^2}{2\delta} = F\alpha\Delta T.$$

The focal length follows the same expansion law as the mirror material.

In practice, non-uniform mirror expansion due to transitional heating or cooling causes higher-order aberrations, in addition to the focus shift. Due to this reason, as uniform cooling may take hours, large telescope mirrors are kept at uniform temperature as much as possible. Also, high thermal conductivity and low heat capacity and low material density are essential for minimizing the thermal aberrations of mirror optics.

2.18 Optical system similar to shown in Fig. 2.10 is used to image infinitely remote object. The filter is positioned 5 cm in front of the lens and has diameter of 1 cm, the lens has diameter of 2 cm and focal length of 5 cm. Determine the positions of the entrance and exit pupils. Derive the formula for the vignetting coefficient as a function of angular field φ.

Solution (2.18). The entrance pupil is coincident with the filter because it is the smallest aperture visible from infinitely remote point on the optical axis. The exit pupil lays at $-\infty$. The vignetting coefficient:

$$\begin{aligned} V = 0 \ \text{ if } \ \varphi &\leq \arctan(0.1) \\ V = 5\tan(\varphi) - 0.5 \ \text{ if } \ \arctan(0.1) < \varphi &< \arctan(0.3) \\ V = 1 \ \text{ if } \ \varphi &\geq \arctan(0.3) \end{aligned}$$

2.19 The radius of planet Niburu is R. The refractive index of the atmosphere of Niburu is described by the function $n = n_0 - h\alpha$ where n_0 is the refraction index on the surface. Find at which height h the light ray sent horizontally will make a complete circle around the planet and return to the original point.[4]

Solution (2.19). The total optical path L around the planet is given by

$$L = 2\pi(R + h)(n_0 - h\alpha).$$

The extremum path satisfy the condition

$$\frac{dL}{dh} = 0,$$

or

$$\frac{dL}{dh} = 2\,(n_0 - \alpha\,h)\,\pi - 2\alpha\,(h + R)\,\pi = 0,$$

Then

$$h = \frac{n_0 - R\,\alpha}{2\,\alpha}.$$

Apparently, the meaningful solution exists only if $n_0 > R\alpha$.

2.20 Telescope with focal length of $F = 1$ m is pointed to a star positioned at the celestial equator. The image sensing is done with a 1x1 cm 1 Mp sensor with square pixels. Calculate the maximum exposure time to limit the image smear to 1 pixel.

[4] Problem F1212 Kvant magazine **1**, p39, (1990, in Russian).

Solution (2.20). The equatorial sky makes full 2π rad rotation in 24 h, corresponding to angular speed of $\nu = 2\pi/24/60/60 \approx 7.3 \times 10^{-5}$ rad/s. The image is moved on the detector with linear speed of $V = F\nu = 7.3 \times 10^{-5}$ m/s. The image sensor with 1 Mp has 1000×1000 square pixels in $1\,\text{cm}^2$, then the pixel pitch is $P = 10^{-5}$ m. Maximum exposure, to limit the smear to one pixel, is

$$\tau = \frac{P}{V} = \frac{10^{-5}}{7.3 \cdot 10^{-5}} = 1.37\,\text{s}.$$

The exposure should be shorter than 1.37 s.

References

1. J.P.C. Southall, *Mirrors, Prisms and Lenses: A Text-book of Geometrical Optics* (Dover Publications, 1964)
2. Y.G. Kozhevnikov, *Optical Prisms (in Russian)* (Mashinostroenie (Moscow), 1984)

Chapter 3
Invariants and Matrices

Abstract The Abbe invariant [1] defines the relationship between the input and output wavefront curvatures after refraction at a spherical surface. For the propagation of conical light pencils, it serves a role analogous to that of Snell's law for individual light rays. The Lagrange invariant [2], conserved throughout an optical system, establishes a relationship between the pupil size, the field of view, and the refractive indices in the object and image spaces. Subsequently, the paraxial model is applied to derive expressions for the depth of focus and the positions of the principal planes in both a thick lens and a two-component optical system. Finally, ray transfer matrices are introduced as a universal method for analyzing paraxial optical systems and Gaussian light beams.

3.1 Abbe Invariant

Consider a divergent spherical wave of radius S propagating in a medium with refractive index n, refracted at a spherical surface of radius R separating into a medium with refractive index n', as illustrated in Fig. 3.1. The propagation time from point A to point B in the first medium must equal the propagation time from point C to point D in the second medium, as shown in the figure.

Equating these propagation times gives

$$\frac{n}{c}\left[\frac{y^2}{2R} - \frac{y^2}{2S}\right] = \frac{n'}{c}\left[\frac{y^2}{2R} - \frac{y^2}{2S'}\right], \tag{3.1}$$

where y is an arbitrary transverse distance chosen for the locations of points A and B, and c is the speed of light. According to the adopted sign convention, $S < 0$, while $R > 0$ and $S' > 0$. Finally, we obtain

$$n\left[\frac{1}{R} - \frac{1}{S}\right] = n'\left[\frac{1}{R} - \frac{1}{S'}\right] = Q, \tag{3.2}$$

where Q is the (null) *Abbe invariant*, characterizing the quadratic phase correction introduced by the refracting surface.

G. Vdovin, *Elementary Technical Optics*, UNITEXT for Physics,
https://doi.org/10.1007/978-3-032-08626-6_3

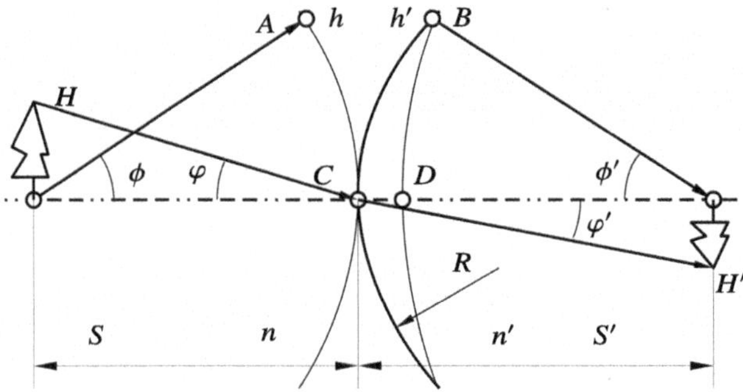

Fig. 3.1 To the derivation of optical invariants

Alternatively, the radius of curvature S' of the refracted wave can be derived from the requirement that the quadratic phase correction, defined by Eq. 2.11, must be continuous across the interface. Equating the phases gives

$$n\frac{2\pi}{\lambda}\left[\frac{y^2}{2R}-\frac{y^2}{2S}\right]=n'\frac{2\pi}{\lambda}\left[\frac{y^2}{2R}-\frac{y^2}{2S'}\right] \quad \text{or}$$
$$n\left[\frac{1}{R}-\frac{1}{S}\right]=n'\left[\frac{1}{R}-\frac{1}{S'}\right]=Q. \tag{3.3}$$

Equation 3.2 can be solved for S' as

$$S'=\frac{RSn'}{(R-S)n+Sn'}. \tag{3.4}$$

In the special case $R=\infty$, we obtain $S'=(n'/n)S$.

Expression (3.4) can also be applied to spherical mirrors by assuming $n'=-1$. In this case, the sign of S' must be inverted after reflection from each mirror surface.

Any paraxial optical system may be represented as a sequence of N spherical surfaces indexed by $i=1,\ldots,N$. Each surface is characterized by radius R_i, refractive index n_i, and spacing t_i. Equation 3.4 can then be used to trace the radius of curvature S through the system defined by $\{R_i, n_i, t_i\}_{i=1}^{N}$, where R_i is the radius of curvature of the i-th surface, n_i is the refractive index of the medium between surfaces i and $i+1$, and t_i is the distance between these surfaces:

$$S_i=\frac{n_i S_{i-1} R_i}{S_{i-1}(n_i-n_{i-1})+n_i R_i}-t_i\ . \tag{3.5}$$

To trace a ray pencil through the system, one starts with the object position S_0 in the medium of refractive index n_0, then applies Eq. 3.5 successively across all surfaces, yielding the final image position S_{N+1} in the medium of refractive index n_{N+1}. For a system in air we assume $n_0=n_{N+1}=1$.

Thus, there is a close analogy between Snell's law–which traces the refraction angles of individual rays at planar interfaces–and the Abbe invariant, which traces the curvature of wavefronts (or ray pencils) through spherical surfaces in the paraxial approximation.

3.2 Lagrange Invariant

Certain relations between the parameters of ray bundles remain unchanged during propagation through the entire optical system. These conserved quantities are extremely useful at the stage of system composition and preliminary design. Typically, such relations hold in the paraxial approximation, where angles are assumed to be infinitely small, so that $\sin\psi$ and $\tan\psi$ can be replaced by ψ, and a spherical surface $R - \sqrt{R^2 + y^2}$ can be replaced by the parabola $y^2/2R$. In essence, this approximation reduces the analysis of a complete optical system to the study of an infinitely thin cylinder cut around its optical axis.

Consider the optical system shown in Fig. 3.1. An object of size H is imaged as H'. For the chief ray passing through the principal point C, Snell's law gives:

$$n\frac{H}{S} = n'\frac{H'}{S'}. \tag{3.6}$$

Next, consider the marginal ray. The height of this ray h (the distance from point A to the optical axis) is preserved in the image space (distance from point B to the axis). Substituting h into both sides of Eq. 3.6, we obtain:

$$nh\frac{H}{S} = n'h\frac{H'}{S'}. \tag{3.7}$$

Finally, we arrive at the compact expression:

$$Hn\psi = H'n'\psi' = \text{const.} \tag{3.8}$$

Equation 3.8 represents one form of the *Lagrange invariant*, which remains constant as light propagates through the system.

For two fundamental rays in an axially symmetrical optical system—the marginal ray (h, φ), originating from the object point on the optical axis and passing through the edge of the entrance pupil, and the chief ray $(\widetilde{h}, \widetilde{\varphi})$, originating from the edge of the object and crossing the nodal point at the pupil center—the invariant takes the form:

$$L = nh\widetilde{\varphi} - n\widetilde{h}\varphi = \text{const.} \tag{3.9}$$

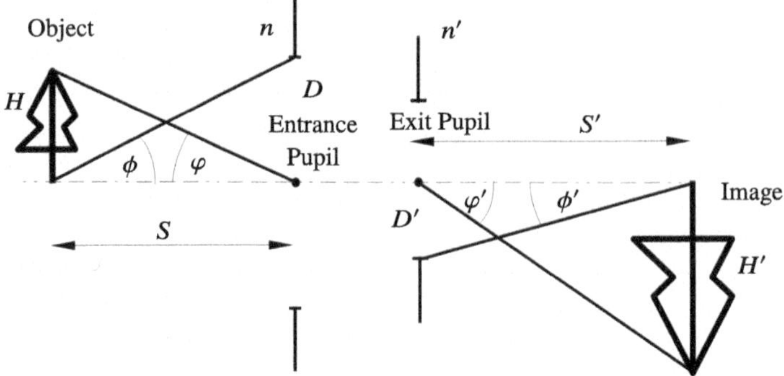

Fig. 3.2 Illustration of the Lagrange invariant

In the object plane, the Lagrange invariant equals the product of the refractive index, object size, and marginal ray angle. In the pupil plane, it equals the product of the refractive index, pupil size, and chief ray angle.

If the distance between the object and the entrance pupil is S, and between the exit pupil and the image is S', the invariant can be written as:

$$n\frac{Hh}{S} = n'\frac{H'h'}{S'} = nH\phi = n'H'\phi' = nh\varphi = n'h'\varphi'. \tag{3.10}$$

This relation demonstrates that, even though the marginal ray height can vary between different pupil images, the invariant is preserved. In a broader sense, Eq. 3.10 provides a general input–output relation for an imaging system, without requiring knowledge of its detailed internal configuration (Fig. 3.2).

The invariant can be evaluated in different ways:

- *Object space*: product of object size, aperture angle, and refractive index of the medium; equivalently, entrance pupil radius times field angle times refractive index.
- *Image space*: product of image size, aperture angle, and refractive index of the medium; equivalently, exit pupil radius times field angle times refractive index.

The Lagrange invariant is related to the information capacity of the system. In the paraxial approximation, the diffraction-limited resolution R in the object plane is:

$$R \simeq \frac{\lambda}{nD}. \tag{3.11}$$

The number of resolved picture elements N can then be expressed as:

$$N = 2\frac{\varphi}{R} = 2\frac{\varphi n D}{\lambda} = 4\frac{L}{\lambda}, \tag{3.12}$$

where L is the Lagrange invariant defined in Eq. 3.9. Thus, N remains constant in all image planes throughout the system. For a two-dimensional image, the total number of resolved pixels equals the product of the resolutions in X and Y. For a symmetrical system, the information capacity is proportional to the square of the Lagrange invariant and inversely proportional to the square of the wavelength. This leads to the concept of *etendue*:

$$E = n^2 dS d\theta, \tag{3.13}$$

where dS is the cross-sectional area of the beam and $d\theta$ is the solid angle[1] subtended by the ray bundle. For two-dimensional images, the information capacity of the system is proportional to the etendue.

In laser optics, etendue is closely related to the beam parameter product (BPP) B, and to the M^2*parameter*, both of which quantify beam divergence. The BPP is defined as the product of the beam waist $w = D/2$ (with D the beam diameter) and the divergence angle φ:

$$B = \varphi w.$$

The M^2 parameter is the BPP normalized to the diffraction limit of a Gaussian beam (described in more detail in Sect. 3.6):

$$M^2 = \frac{\pi w \varphi}{\lambda} = \frac{\pi B}{\lambda} = \frac{\varphi}{\varphi_0},$$

where φ_0 is the divergence of an ideal Gaussian beam with waist w and wavelength λ. Thus, $M^2 = 1$ for a perfect Gaussian beam. In practice, M^2 indicates how much larger the focused spot of a real beam is compared to the diffraction-limited Gaussian spot. High-quality single-mode lasers achieve $M^2 < 1.5$, while high-power multimode lasers may have M^2 values ranging from 10 to 100.

3.3 Thick Lens: Principal Planes

In a real lens of thickness D and refractive index n, bounded by two spherical surfaces of radii R_1 and R_2, the single principal plane of a thin lens splits into two distinct principal planes, which are conjugated to each other. The optical behavior of such a thick lens is fully determined by its focal length together with the positions of its principal planes.

By definition, a ray originating at the focal point enters the first principal plane at height H and emerges from the second principal plane at the same height H, parallel to the optical axis.

The positions of the principal planes can be derived from the geometry in Fig. 3.3. Applying the Abbe invariant to rays emerging from the first focal point gives the

[1] $d\theta \approx \pi\varphi^2$, where φ is the divergence angle of the beam.

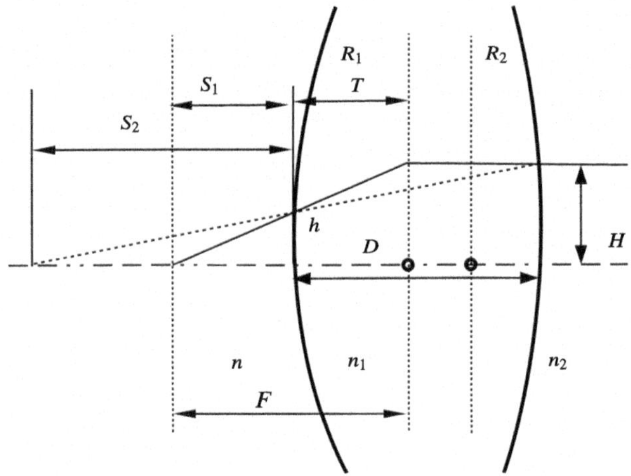

Fig. 3.3 Principal planes of a thick lens

system of equations:

$$n\left(\frac{1}{R_1}-\frac{1}{S_1}\right)=n_1\left(\frac{1}{R_1}-\frac{1}{S_2}\right),\tag{3.14}$$
$$n_1\left(\frac{1}{R_2}-\frac{1}{S_2-D}\right)=\frac{n_2}{R_2}.$$

For the common case of a lens in air ($n=n_2=1$), solving Eq. 3.14 yields

$$S_1=-\frac{(R_1R_2+DR_1)n-DR_1}{(R_2-R_1+D)n^2+(-R_2+R_1-2D)n+D},\qquad S_2=\frac{(R_2+D)n-D}{n-1}.\tag{3.15}$$

To determine T, note that the height H is given by $H=(-h/S_1)(T-S_1)=(-h/S_2)(D-S_2)$, which implies $S_1D=S_2T$, hence $T=DS_1/S_2$. For a lens in air this becomes

$$T=\frac{DR_1}{(R_1-R_2-D)n_1+D}.\tag{3.16}$$

The position of the back principal plane relative to the first surface of the lens is then found by applying Eq. 3.16 to the reversed lens:

$$T'=D+\frac{DR_2}{(R_1-R_2-D)n_1+D}.\tag{3.17}$$

Finally, the focal length of a thick lens is given as the distance from the focal point to the front surface (S_1, negative) plus the position of the first principal plane T:

$$\begin{aligned} F = T - S_1 &= \frac{R_1 R_2 n_1}{(R_2 - R_1 + D)n_1^2 + (-R_2 + R_1 - 2D)n_1 + D} \\ &= \left[(n_1 - 1)\left\{ \frac{1}{R_1} - \frac{1}{R_2} + \frac{D(n_1 - 1)}{R_1 R_2 n_1} \right\} \right]^{-1}. \end{aligned} \tag{3.18}$$

This relation is the well-known lensmaker's equation, connecting lens geometry with focal length F.

The locations of the principal planes depend strongly on the lens shape. In the special case of a plano-convex or plano-concave lens of thickness D and refractive index n_1, the first principal plane coincides with the apex of the curved surface, while the second lies inside the lens at a distance D/n_1 from the flat surface. The separation between the principal planes is therefore

$$D - T - T' = \frac{n_1 - 1}{n_1} D. \tag{3.19}$$

This result coincides with the focus shift introduced by a plane-parallel plate, since in the paraxial approximation a plano-convex lens can be represented as the combination of a thin spherical refracting surface (providing optical power) and a plane-parallel plate (representing the lens bulk).

3.4 Ray Matrices

Successive application of the Abbe invariant to a sequence of optical surfaces allows one to trace a ray pencil, but the resulting expressions often become cumbersome. An alternative method, mathematically equivalent to the application of Abbe invariant, is based on matrix algebra and provides a more compact and convenient description of ray propagation.

The ray matrix formalism is particularly useful for tracing paraxial meridional rays through rotationally symmetric optical systems composed of spherical surfaces. In this approximation, a ray is completely defined by a column vector containing its height h and its reduced angle ψ:

$$\begin{vmatrix} h \\ \psi \end{vmatrix}.$$

Here, the reduced angle is defined as $\psi = n\varphi$, where φ is the geometrical ray angle with respect to the optical axis and n is the refractive index of the medium in which the ray propagates, following the convention of [3].

Propagation through an optical system is then expressed as a linear transformation:

Table 3.1 Elementary paraxial ray matrices [3]

Free space of length L and refractive index n	$\begin{vmatrix} 1 & \frac{L}{n} \\ 0 & 1 \end{vmatrix}$
Flat interface $n_1 \to n_2$ (reduced angles)	$\begin{vmatrix} 1 & 0 \\ 0 & 1 \end{vmatrix}$
Spherical interface $n_1 \to n_2$, radius r	$\begin{vmatrix} 1 & 0 \\ -\frac{n_2 - n_1}{r} & 1 \end{vmatrix}$
Spherical mirror, radius r	$\begin{vmatrix} 1 & 0 \\ -\frac{2}{r} & 1 \end{vmatrix}$
Positive thin lens, focal length F	$\begin{vmatrix} 1 & 0 \\ -\frac{1}{F} & 1 \end{vmatrix}$

$$\begin{aligned} h_1 &= Ah + B\psi, \\ \psi_1 &= Ch + D\psi, \end{aligned} \tag{3.20}$$

or, in compact matrix form,

$$\begin{vmatrix} h_1 \\ \psi_1 \end{vmatrix} = \begin{vmatrix} A & B \\ C & D \end{vmatrix} \begin{vmatrix} h \\ \psi \end{vmatrix}. \tag{3.21}$$

Ray matrices for entire systems are obtained by multiplying the elementary matrices of all components between the input and output planes. If a ray passes sequentially through components C_1, C_2, and C_3, then the overall matrix is written as

$$|C_3| \cdot |C_2| \cdot |C_1|,$$

with the product ordered from right to left. The most common elementary ray matrices are listed in Table 3.1.

Consider now the general ray matrix describing propagation from a plane P_1 in medium of index n_1 to a plane P_2 in medium of index n_2. This matrix is the product of all element matrices between P_1 and P_2. Its elements have the following interpretations:

- The determinant $AD - BC$ is always equal to 1.
- If $D = 0$, then P_1 is the first focal plane of the system.
- If $B = 0$, then P_1 and P_2 are conjugate planes; the coefficient A gives the linear magnification.
- If $C = 0$, the system is afocal (telescopic); the angular magnification is $n_1 D / n_2$.
- If $A = 0$, then P_2 is the back focal plane of the system.

Although matrix optics is handy to describe paraxial optical systems, correct description of real optical systems formed by spherical surfaces requires exact tracing of optical rays. Computationally efficient expressions for exact ray tracing through spherical and aspherical optical surfaces were derived by Feder in 1951 in [4].

3.5 Matrix Description of a General Imaging System

The treatment of a thick lens can be generalized to an arbitrary imaging system composed of multiple spherical components. Figure 3.4 illustrates the cardinal points and principal planes of such a system. Two principal planes, PP_1 and PP_2, are defined; by construction they are conjugate to each other. In addition, the system possesses two nodal points, N_1 and N_2, such that any ray entering through N_1 will emerge from N_2 with the same direction.

Table 3.2 summarizes the principal parameters of the optical system in terms of the elements of its ray matrix, which relates the input plane P_1 to the output plane P_2.

The nodal points coincide with the principal planes when both the input and output planes lie in media with the same refractive index. If the indices differ–as in immersion microscopy or underwater imaging–the nodal points no longer coincide with the principal planes and are shifted relative to them.

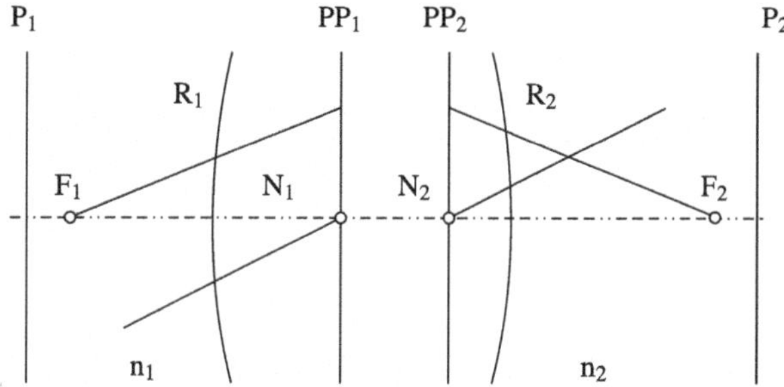

Fig. 3.4 Cardinal points and planes of a general imaging system

Table 3.2 Optical system parameters expressed through the ray matrix elements [3]

Parameter	Measured from → to	Expression
First focus	$P_1 \to F_1$	$n_1 D/C$
First focal length	$PP_1 \to F_1$	$-n_1/C$
First principal point	$P_1 \to PP_1$	$-n_1(D-1)/C$
First nodal point	$P_1 \to N_1$	$-(Dn_1 - n_2)/C$
Second focus	$P_2 \to F_2$	$-n_2 A/C$
Second focal length	$PP_2 \to F_2$	$-n_2/C$
Second principal point	$P_2 \to PP_2$	$n_2(1-A)/C$
Second nodal point	$P_2 \to N_2$	$(n_1 - An_2)/C$

3.6 Gaussian Beams

Gaussian beams represent a special class of optical beams with unique and practically important properties. The key feature is that the functional form describing their intensity distribution remains unchanged upon propagation. In other words, the transverse field distribution of a Gaussian beam is an eigenfunction of the free-space propagation operator. Only the beam width w and the wavefront curvature ρ evolve with distance.

A Gaussian beam is fully described by its complex radius of curvature $\widetilde{\rho}$ that combines the beam width w and the radius of curvature ρ:

$$E_0 \exp\left(-\frac{x^2+y^2}{w^2}\right)\exp\left(\frac{\mathrm{i}k(x^2+y^2)}{2\rho}\right) = E_0 \exp\left(\frac{\mathrm{i}k(x^2+y^2)}{2\widetilde{\rho}}\right), \tag{3.22}$$

where $\widetilde{\rho}$ is defined as

$$\frac{1}{\widetilde{\rho}} = \frac{1}{\rho} + \frac{\mathrm{i}\lambda}{\pi w^2}. \tag{3.23}$$

Here, w is the beam radius at the $1/e$ amplitude level, ρ is the geometric radius of curvature of the wavefront, and E_0 is the field amplitude. Since intensity is proportional to the squared field, the $1/e^2$ intensity fall occurs at $r = w$, enclosing about 85% of the beam power. The radius at half-maximum intensity is approximately $0.6w$ and contains 50% of the total power.

In a uniform isotropic medium, the Gaussian beam is symmetric with respect to its waist. At the waist, the beam radius is w_0 and the wavefront is flat ($\rho = \infty$).

The confocal parameter is defined as

$$z_c = \frac{\pi w_0^2}{\lambda}, \tag{3.24}$$

and represents the distance over which the beam radius expands by a factor of $\sqrt{2}$ compared to the waist.

Starting from the waist, the beam radius w and curvature ρ evolve with propagation distance z as

$$w(z) = w_0\sqrt{1+\left(\frac{z}{z_c}\right)^2},$$
$$\rho(z) = z\sqrt{1+\left(\frac{z_c}{z}\right)^2}, \tag{3.25}$$

with $w(0) = w_0$.

In the more general case, the complex radius of curvature $\widetilde{\rho}$ can be propagated through any paraxial system using the so-called *ABCD rule*. For a wavefront with initial curvature R_1 passing through a system with ray matrix elements A, B, C, D, the output curvature R_2 is

$$R_2 = \frac{AR_1 + B}{CR_1 + D}. \tag{3.26}$$

For Gaussian beams described by Eq. 3.22, the ABCD rule applies directly to the complex radius of curvature $\widetilde{\rho}$:

$$\widetilde{\rho}_2 = \frac{A\widetilde{\rho}_1 + B}{C\widetilde{\rho}_1 + D}. \tag{3.27}$$

This compact relation enables calculation of the evolution of w and ρ through any paraxial optical system.

Gaussian beams are of central importance in laser optics. They exhibit the smallest possible divergence of all beam profiles and retain their Gaussian form under propagation. In practice, lasers oscillating in the fundamental transverse mode produce beams that are close to Gaussian. However, a truly ideal Gaussian beam would require infinite aperture, and thus cannot be realized experimentally.

3.7 Beam Quality, Diffraction Limit, and Invariants

Beam quality is quantified by the parameter M^2, which compares the real beam to the ideal Gaussian case. A perfect Gaussian beam has $M^2 = 1$. If $M^2 > 1$, the deviation may arise from:

- Non-Gaussian intensity distribution. For example, a uniform circular beam has $M^2 \sim 1.5$ due to diffraction at the aperture edges, even with an ideal wavefront.
- Phase aberrations. The most common reason for elevated M^2. These distort the wavefront and degrade beam quality.

Phase aberrations strongly influence the M^2 parameter, but intensity distribution must always be considered, especially when $M^2 \approx 1$. A beam with non-Gaussian intensity cannot achieve $M^2 = 1$, regardless of its wavefront quality.

The asymptotic divergence angle of a Gaussian beam is

$$\varphi = \frac{\lambda}{\pi w_0} \approx 0.32\frac{\lambda}{w_0},$$

where w_0 is the beam waist radius at the $1/e^2$ intensity level. This divergence is significantly smaller than that of a uniform beam of radius r_0:

$$\varphi \approx 0.61\frac{\lambda}{r_0}.$$

However, the Gaussian profile extends infinitely, so to realize this small divergence, the system aperture must be much larger than w_0. Figure 3.5 shows that when truncated at w_0, a Gaussian beam diverges more strongly than a beam with uniform intensity distribution, of equal power.

The M^2 parameter provides a practical measure of beam quality relative to the ideal Gaussian case. For an ideal Gaussian beam, $M^2 = 1$, and the divergence angle is minimized according to

$$\varphi_{\min} = \frac{\lambda}{\pi w_0}.$$

For a real beam with $M^2 > 1$, the divergence increases proportionally:

$$\varphi = M^2 \frac{\lambda}{\pi w_0}.$$

At the same time, the waist radius w_0 achievable for a given focusing system also scales with M^2. Thus, M^2 directly quantifies the departure from the diffraction-limited performance of an optical system.

This formulation naturally connects to the *diffraction limit* of imaging systems discussed in Sect. 3.1. Diffraction defines the smallest resolvable feature in terms of aperture diameter D and wavelength λ. In Gaussian beam optics, the same physical limitation is defined by the relation between the beam waist w_0 and divergence angle φ. Both viewpoints highlight the fundamental trade-off between spatial confinement and angular spread of light.

The relation between M^2 and the system invariants is particularly significant. The Lagrange invariant, which links object size, numerical aperture, and refractive index, sets the maximum number of spatial modes that an optical system can accept or transmit. In beam optics, this translates to the number of supported transverse modes:

$$N_{\text{modes}} \approx (M^2)^2.$$

Thus, M^2 not only describes beam quality but also provides an estimate of how many higher-order spatial modes are mixed into the beam. A system with $M^2 = 1$ supports only the fundamental Gaussian mode, while $M^2 \gg 1$ indicates the presence of multiple incoherently mixed modes.

From the practical standpoint, this connection has far-reaching implications:

- In laser design, minimizing M^2 ensures that the output beam can be tightly focused, achieving diffraction-limited spot sizes.
- In optical communication, M^2 characterizes the mode content of beams propagating in fibers, directly affecting bandwidth and coupling efficiency.
- In imaging systems, M^2 represents the effective reduction of system resolution relative to the diffraction limit.

Therefore, the M^2 parameter serves as a bridge between the abstract invariants of optical systems and the measurable performance metrics of real beams and instru-

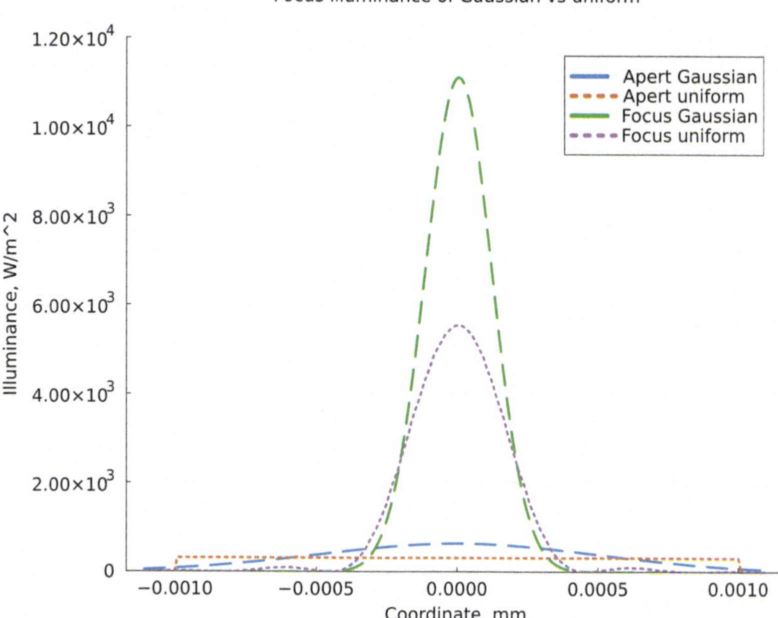

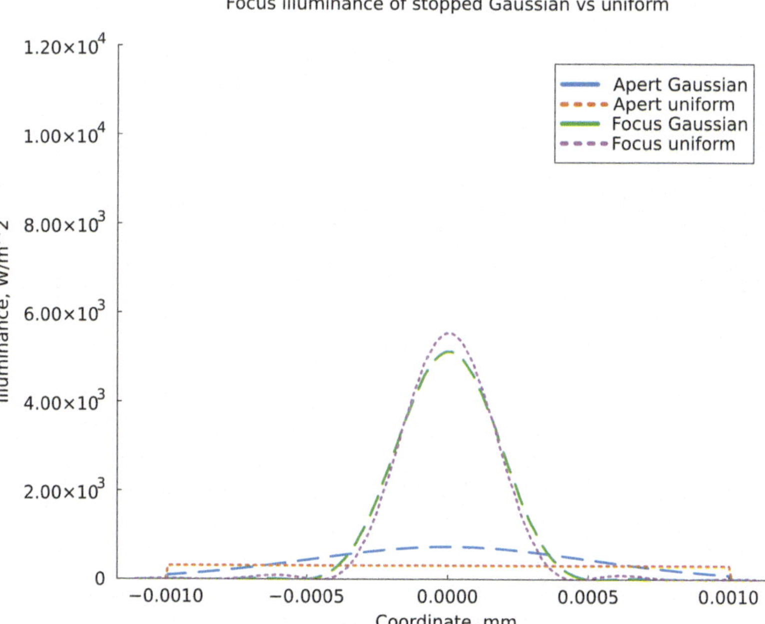

Fig. 3.5 Numerical simulation of focusing unlimited and aperture-limited Gaussian beams of the same power. A 1 mW coherent beam with $\lambda = 1\ \mu$m is focused by a 2 mm aperture ($w_0 = r_0 =$ 1 mm) at a distance of 0.75 m. The upper graph shows that a very large aperture is required to achieve the theoretical small divergence of the Gaussian beam. The lower graph shows that if the beam is truncated at w_0, maintaining total power, then a uniform beam of the same power exhibits superior focusing performance

ments. It encapsulates both the fundamental diffraction constraints and the imperfections introduced by aberrations, misalignment, or multimode excitation.

3.8 Conclusion

The Abbe invariant provides a compact relation for tracing wavefront curvature through spherical interfaces. Its formulation,

$$n\left(\frac{1}{R}-\frac{1}{S}\right)=n'\left(\frac{1}{R}-\frac{1}{S'}\right),$$

links object and image distances S, S' with surface radius R and refractive indices n, n'. Applied sequentially, this relation traces the evolution of curvature in complex systems, paralleling Snell's law for rays. For mirrors, the same expression holds by substituting $n' = -1$ with inversion of signs.

The Lagrange invariant is conserved through the whole optical system. It is directly related to etendue and M^2 beam quality factor. Also the information capacity of an optical system depends on the Lagrange invariant.

In thick lenses, formed by two spherical surfaces with separation D, the focal length is defined as

$$\frac{1}{F}=(n-1)\left(\frac{1}{R_1}-\frac{1}{R_2}+\frac{(n-1)D}{nR_1R_2}\right)$$

expresses the focal length F in terms of radii R_1, R_2, refractive index n, and thickness D. Thick lens has two separate principal planes, position of which depends on the lens thickness and surface radii.

The matrix formalism uses linear algebra to trace ray coordinates through optical system in paraxial approximation. A ray defined by height h and reduced angle $\psi = n\varphi$ transforms via

$$\begin{vmatrix} h_1 \\ \psi_1 \end{vmatrix} = \begin{vmatrix} A & B \\ C & D \end{vmatrix} \begin{vmatrix} h \\ \psi \end{vmatrix},$$

where elementary matrices describe free space, refractions, and lenses. Product of separate component matrices yields the system matrix, from which focal and principal planes, and other system paranmeters, can be directly calculated.

Gaussian beam propagation can be modeled with ABCD matrices. The complex radius of curvature of the Gaussian beam

$$\frac{1}{\widetilde{\rho}}=\frac{1}{\rho}+\frac{i\lambda}{\pi w^2}$$

transforms according to ABCD rule:

$$\widetilde{\rho}_2 = \frac{A\widetilde{\rho}_1 + B}{C\widetilde{\rho}_1 + D}.$$

Thus beam waist w and curvature ρ evolve through arbitrary paraxial systems. The M^2 parameter quantifies departure from the ideal Gaussian.

Invariants and matrix methods provide a unified framework for analyzing paraxial imaging and propagation of Gaussian beams under diffraction-limited conditions.

3.9 Problems

3.1 Derive expressions for the focal length and the principal plane positions of the

1. Ball lens with radius R and refraction index n.
2. Meniscus lens formed by two concentric surfaces with radii R_1 and R_2.

Solution (3.1). Ball lens: $R_2 = R = -R_1$ and $D = 2R$. Substituting into the lens-maker's equation:

$$F = ((n-1)(2/R - (n-1)2R/n/R^2))^{-1} = \frac{Rn}{2(n-1)}$$

The position of the principal plane is given by Eqs. (3.8)–(3.10), substituting the ball data we obtain $T = R$. Then, since the ball is symmetrical, $T' = R$. The principal planes of the ball lens are co-incident and cross the ball in the middle.

Meniscus: Substituting $R_2 = R_1 - D$ into Eqs. (3.8)–(3.10) we obtain:

$$F = -\frac{\left(R_1^2 - D\,R_1\right)\,n}{D\,n - D},$$

and $T = T' = R_1$. The principal planes are coincident and pass through the center of curvature.

3.2 Using the Abbe invariant find the radius R of a glass ball with refractive index n, providing the infinitely remote object is imaged on the back surface of the ball.

Solution (3.2). Assuming $S = \infty$ and directly applying the Abbe invariant (3.2) we equate:

$$\frac{1}{R} - \frac{1}{\infty} = n[\frac{1}{R} - \frac{1}{2R}], \tag{3.28}$$

of which the only solution is: $n = 2$. The solution does not depend on the ball radius R. Any transparent ball with $n = 2$ focuses infinitely remote object on its back surface. Obviously, this solution is valid only in the paraxial area.

3.3 Explain why objects seem closer for an observer wearing diving mask underwater. Estimate the visible distance shift, using paraxial approximation.

Solution (3.3). After refraction on the water-air interface, light cones enter the human eye with an apex angle that is $\sim n$ times larger than they would enter without refraction, where $n \approx 1.33$ for water. Then, all objects have 1.33 times larger angular size, and look 1.33 times closer to the observer.

3.4 Plano-convex lens with thickness $D = 5$ mm made of glass with refractive index of $n = 1.5$ produces image with magnification $M = -2$. The distance between the object and the image is $L = 100$ mm. Calculate the radius of curvature R of the lens surface, the distance from the front surface of the lens to the object S, and the distance from the back surface of the lens to the image S'. Assume the lens is turned to the image with its curved surface.

Solution (3.4). Let a be the distance from the front principal plane to the object, and b be the distance from the back principal plane to the image. Then, assuming both distances are positive, and taking into account the distance between principal planes defined by Eq. 3.19:

$$\frac{b}{a} = |M|,$$
$$L = a + b + \frac{n-1}{n} D, \tag{3.29}$$

from which we find:

$$a = \frac{L - D(n-1)/n}{|M| + 1} \tag{3.30}$$
$$b = |M| a, \tag{3.31}$$

and, using the thin lens equation

$$F = (a^{-1} + b^{-1})^{-1}$$
$$R = \frac{F}{n-1}$$
$$S' = b$$
$$S = L - D - S' \tag{3.32}$$

Substituting our data, we have: $R = 10.926$ mm convex, $S' = b = 65.555$, $S = 29.444$ mm.

3.5 To measure the focal length of the imaging objective, and the positions of its principal planes, the objective is placed between two conjugated planes, separated by distance L, producing sharp image with magnification M, satisfying to the condition $|M| > 1$. Then the optical system is shifted by distance l to produce sharp image with magnification M^{-1}. Determine the focal length F of the optical system, the positions of its principal planes and the separation Δ between the principal planes.

Solution (3.5). Let a be the distance from the front principal plane to the object, and b is the distance from the back principal plane to the image. Then, we can state:

$$\frac{b}{a} = |M|$$
$$\frac{b-l}{a+l} = |M|^{-1},$$

with a solution

$$a = \frac{l}{|M|-1}$$
$$b = \frac{|M|l}{|M|-1}. \tag{3.33}$$

The focal length is found as

$$F = \frac{ab}{a+b} = \frac{|M|l}{M^2-1},$$

the first principal plane is positioned at distance a from the object, and the separation between two principal planes is given by $\Delta = L - a - b$.

3.6 Using Abbe invariant make Excel spreadsheet that traces the radius of curvature of a paraxial light pencil with radius of curvature S through a sequence of optical surfaces defined by the parameters $R,\ d,\ n$, where R is the radius of curvature of the surface, d is the distance to the next surface, n is the index of refraction.

3.7 Using Expr. (3.16), (3.17), (3.18) sketch the positions of principal planes and the front/back focal points for lenses with $D = 1, n = 1.5$:
$R_1 = 10,\ R_2 = \infty$
$R_1 = -10,\ R_2 = \infty$
$R_1 = 10,\ R_2 = -10$
$R_1 = -10,\ R_2 = 10$
$R_1 = 10,\ R_2 = 10$
$R_1 = 10,\ R_2 = 9$
$R_1 = -10,\ R_2 = -10$

3.8 The entrance pupil of a photographic objective is two times larger than the exit pupil. Carla's height H is 1 m 15 cm tall. She stays at a distance of $L = 3$ m in front of the entrance pupil. The image of Carla is obtained at a distance of $L' = 5$ cm behind the exit pupil. Calculate the size H' of Carla's image.

Solution (3.8). Using Lagrange invariant we state:

$$D\frac{H}{L} = \frac{D}{2}\frac{H'}{L'},$$

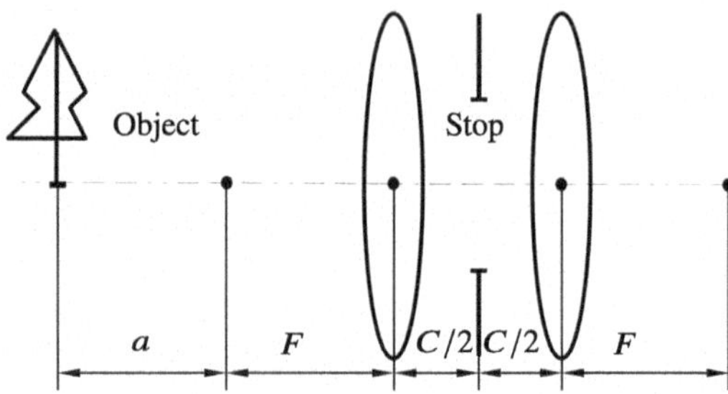

Fig. 3.6 To Problem 3.9

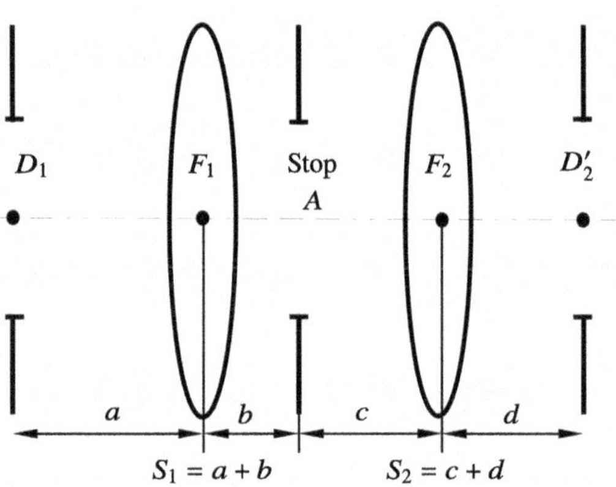

Fig. 3.7 To Problem 3.10

where D is the diameter of the entrance pupil. Then

$$H' = \frac{2L'H}{L} = 0.03833\ldots \text{ m}.$$

3.9 System shown in Fig. 3.6 is formed by two positive lenses, each with focal length F, separated by distance C. The stop with diameter D is positioned in the middle between the lenses. Calculate sizes and positions of the entrance and exit pupils, trace chief and marginal rays through the system for an object at distance a with:

- $C = F/2, \quad a = F;$
- $C = F, \quad a = F;$
- $C = 2F, \quad a = \infty;$
- $C = 2F, \quad a = 0;$

3.10 The optical system shown in Fig. 3.7 consists of 2 thin lenses with aperture diaphragm of diameter A positioned between them. The entrance pupil has diameter D_1 and positioned at a distance S_1 in front of the diaphragm. The exit pupil has diameter D_2 and is positioned at a distance S_2 behind the diaphragm. Calculate the focal length of the system.

Solution (3.10). The focal length F of the system shown in Fig. 3.7 satisfies to

$$\frac{1}{F} = \frac{1}{F_1} + \frac{1}{F_2} - \frac{b+c}{F_1 F_2}. \tag{3.34}$$

The values of a, b, and F_1 are defined by the system of equations:

$$\begin{aligned} a + b &= S_1 \\ D_1/A &= \pm a/b \\ a^{-1} \pm b^{-1} &= F_1^{-1}, \end{aligned}$$

with a solution (only for "+" sign in the above equations with $\pm$):

$$\begin{aligned} b &= \frac{A\, S_1}{D_1 + A} \\ F_1 &= \frac{A\, D_1\, S_1}{(D_1 + A)^2}. \end{aligned}$$

In a similar way we find:

$$\begin{aligned} c &= \frac{A\, S_2}{D_2 + A} \\ F_2 &= \frac{A\, D_2\, S_2}{(D_2 + A)^2}. \end{aligned} \tag{3.35}$$

The focal length is, using (3.34):

$$F = \frac{D_1 D_2 S_1 S_2}{(D_1 + A)^2 S_2 + (D_2 + A)^2 S_1}. \tag{3.36}$$

3.11 Reflective light sail with radius of $r = 100$ m, reflectivity $R = 0.9$, and area density of $\rho = 1$ g/m^2 is illuminated by a laser placed on the Earth surface. The laser provides the sail with acceleration of $a = 0.1$ m/s^2. Estimate the laser beam power.
Solution (3.11). The pressure P is defined by the second law of Newton as

$$p = \rho a.$$

The same pressure is defined through illuminance I [W/m^2] as

$$p = (1 + R)\frac{I}{c}.$$

Finally, for the total power P of the light beam we obtain:

$$P = \pi r^2 I = \frac{\pi r^2 \rho a c}{1 + R} = 5 \cdot 10^8 \text{ W}.$$

3.12 Reflective light sail has radius of $r = 100$ m. It is illuminated from Earth, at a distance of $L = 100 \cdot 10^6$ km. Assuming the illumination is Gaussian, the beam radius on the level of $1/e^2$ by intensity is equal to the sail radius, the wavelength is $\lambda = 1\,\mu$m, estimate the radius R of the illumination telescope on Earth.

Solution (3.12). The confocal parameter:

$$z_c = \frac{\pi r^2}{\lambda} \approx 3.14 \cdot 10^{10},$$

For the beam radius on Earth we can use the property of symmetry of the Gaussian beam around its waist. Since the beam waist is on the sail, the beam radius back on Earth is the same as if the beam has propagated the same distance forward:

$$R = r\sqrt{1 + \left[\frac{L}{z_c}\right]^2} \approx 333 \text{ m}.$$

To preserve the low divergence of the Gaussian beam, as illustrated by Fig. 3.5, the telescope aperture should be about twice the calculated radius. So, to drive the sail, the telescope diameter on Earth should be larger than 1 km.

3.13 In a space communication system, Gaussian beam with waist of $w_0 = 5$ cm and $\lambda = 1.55\,\mu$m is focused to the receiver station at a distance of 100 km. Estimate the necessary angular pointing precision for the transmitter. Estimate the relative loss of power in the receiver, if the transmitted beam experiences defocus of 0.1 μm, 1 μm, 2 μm, 5 μm. Define the tolerances for pointing and focusing mechanisms. Explain the assumptions made.

Solution (3.13). We will use the ABCD rule

$$\rho_1 = \frac{A\rho + B}{C\rho + D}$$

where A, B, C, D are the elements of ray matrix, describing the optical system, and ρ is the complex curvature of the Gaussian beam with radius of curvature R and waist $w = w_0$:

$$\frac{1}{\rho} = \frac{1}{R} + \frac{i\lambda}{\pi w^2}, \quad \rho = \frac{\pi R w^2}{\pi w^2 + i R\lambda}.$$

Initially $R = \infty$, but we keep it in the formulas, as in the future it will be varied. The matrix for propagation to distance L would be

$$\begin{vmatrix} A & B \\ C & D \end{vmatrix} = \begin{vmatrix} 1 & L \\ 0 & 1 \end{vmatrix}.$$

Then, applying the ABCD rule, we get:

$$\rho_1 = \frac{\pi R w^2}{\pi w^2 + iR\lambda} + L.$$

and noting that

$$\Re\left(\frac{1}{\rho_1}\right) = \frac{1}{R_1}, \text{ and } \Im\left(\frac{1}{\rho_1}\right) = \frac{i\lambda}{\pi w_1^2}$$

where $\Re$ and $\Im$ denote the real and imaginary parts of the expression, we find:

$$R_1 = \frac{L^2 R^2 \lambda^2 + (R+L)^2 \pi^2 w^4}{L R^2 \lambda^2 + (R+L)\pi^2 w^4}$$

$$w_1^2 = \frac{L^2 R^2 \lambda^2 + (R+L)^2 \pi^2 w^4}{R^2 \pi^2 w^2}.$$

By differentiating the expression for w^2 with respect to R and equating to zero we find that the condition

$$R = -L$$

corresponds to the minimum of w_1^2. Practically it means that, for the best transmission, the beam should be focused to the target. For the best focus we have

$$w_{1best} = \frac{L\lambda}{\pi w} = \frac{10^5 \cdot 1.55 \cdot 10^{-6}}{\pi \cdot 5 \cdot 10^{-2}} \approx 1 \text{ m}.$$

Apparently, the pointing precision should be much better than $w_{1best}/L \sim 10^{-5}$ rad. 10^{-6} rad would be a good initial guess.

For the value of w_1 as a function of L and defocus aberration $\Delta = w^2/2/R$ we obtain:

$$w_1 = \frac{\sqrt{L^2\lambda^2 + \pi^2 w^4 + 4\Delta L S \pi^2 w^2 + 4\Delta^2 L^2 \pi^2}}{\pi w}.$$

The axial intensity $I \sim w_{1best}^2/(w_1(\Delta)^2)$ as a function of defocus Δ is illustrated by Fig. 3.8. Apparently the maximum defocus aberration should not exceed 100 nm, in good agreement with the $\lambda/12$ quality criteria.

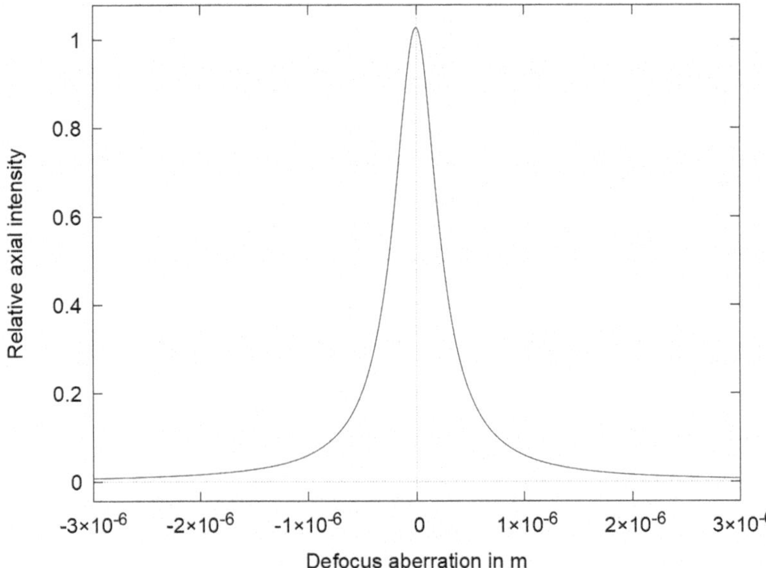

Fig. 3.8 Axial intrensity in the receiving satellite as a function of defocus

References

1. W.J. Smith, *Modern Optical Engineering 4E (PB)* (McGraw Hill LLC, 2007)
2. J.E. Greivenkamp, *Field Guide to Geometrical Optics*. Field Guides (SPIE Press, 2004). ISBN: 9780819452948
3. A. Gerrard, J.M. Burch, *Introduction to Matrix Methods in Optics* (Dover Publications, Incorporated, 2012)
4. D.P. Feder, Optical calculations with automatic computing machinery. J. Opt. Soc. Am. **41**(9), 630–635 (1951)

Chapter 4
Aberrations of Optical Systems

Abstract Optical aberrations are defined as deviations from ideal wavefront sphericity or deviations from the homocentricity of a ray bundle. Aberrations are inherent in any optical system composed of spherical surfaces. Aberration polynomials are introduced by decomposing an aberrated wavefront into a Taylor series. This decomposition reveals that low-order wavefront aberrations–namely, defocus, coma, astigmatism, and spherical aberration, represented by Seidel polynomials depending on pupil coordinates and field angle—have the highest magnitude. The correction of these low-order aberrations is of primary importance and can be achieved through the proper design of optical systems utilizing multiple refractive and/or reflective surfaces. Zernike polynomials are introduced as an orthogonal set for representing wavefront error across the pupil. The concepts of aplanatic points on spherical surfaces and the use of aspheric surfaces described by conic sections (both refractive and reflective) are presented. Glass dispersion is the primary source of chromatic aberration, which can be reduced by using appropriate combinations of different glass types in the optical design. Finally, concepts of point spread function (PSF) and the modulation transfer function (MTF) are introduced for proper analysis of aberrations influence on the image quality.

4.1 Geometrical Aberrations

A thorough treatment of aberration theory is given in [1, 2]. Here we summarize the concepts of greatest practical relevance.

The ideal image of a point source is formed by a ray bundle that converges precisely to the image point. Such an ideal bundle has a perfectly spherical wavefront W_i. In reality, however, many factors distort the wavefront, leading to image smear and loss of sharpness. The actual, aberrated wavefront W_a can be expressed as the sum of the ideal wavefront W_i and the aberration function W, i.e. $W_a = W + W_i$. The *wavefront aberration* function $W(x, y)$ is defined over the pupil of the optical system. If a ray passes through the pupil at coordinates (x, y), its transverse deviation from the ideal image point in the image plane can be written as

G. Vdovin, *Elementary Technical Optics*, UNITEXT for Physics,
https://doi.org/10.1007/978-3-032-08626-6_4

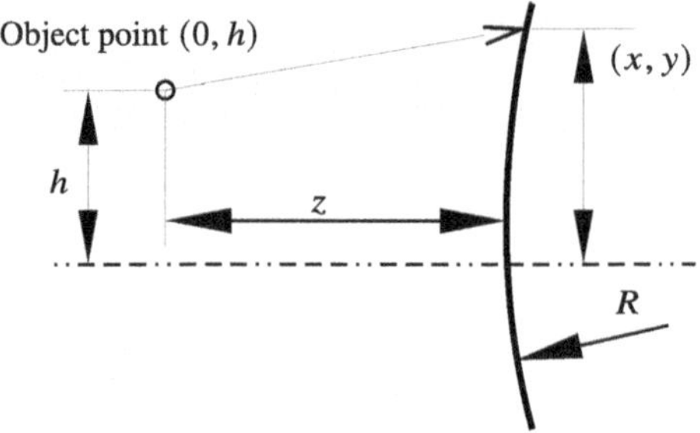

Fig. 4.1 Geometry illustrating the derivation of expressions for third-order aberrations

$$\Delta x = \frac{dW}{dx} S'$$
$$\Delta y = \frac{dW}{dy} S', \qquad (4.1)$$

where S' is the distance from the exit pupil to the image plane, and the derivatives are calculated in points where ray intersects with the pupil.

Two distinct functions of transverse aberrations are usually defined across the pupil: meridional and sagittal. From Eq. 4.1 it follows that the polynomial order of the transverse aberration is by one lower than that of the wavefront aberration. Thus, if wavefront aberrations are expressed as polynomials up to the 4th order, the corresponding transverse aberrations are of the 3rd order. These third-order aberrations are particularly important in system design because they can be derived analytically by paraxial ray tracing.

The wavefront aberration W provides a bridge between diffraction theory and geometrical optics. Within scalar diffraction theory, it determines the focal distribution by introducing a phase shift kW, where $k = 2\pi/\lambda$ is the wave number. At the same time, it allows the derivation of transverse aberrations through geometrical considerations. Importantly, W depends not only on pupil coordinates (x, y) but also on the position of the object point (h, z) relative to the system. Thus, wavefront aberrations vary with both object height h and object distance z.

Figure 4.1 illustrates the geometry of a simple case: a single spherical surface of radius R, illuminated by an off-axis object point at $(0, h)$ located a distance z away. Owing to axial symmetry, the x coordinate of the object can be set to zero without loss of generality. The wavefront will be aberration-free if it coincides with the spherical surface, independent of whether the surface is refractive or reflective. For all other situations, the aberration function W can be approximated as the scaled difference between the optical surface and the inclined spherical wavefront emitted by the object point:

$$W = z - \sqrt{z^2 - (y-h)^2 - x^2} + R - \sqrt{R^2 - x^2 - y^2}. \qquad (4.2)$$

Table 4.1 Third-order wavefront aberrations expressed in terms of pupil coordinates and field angle

Name	Expression	Properties
Spherical aberration	$C_{sa}(x^2+y^2)^2$	Fourth order in pupil
Coma	$C_{coma1}\,\alpha y(x^2+y^2)$	Linear in field, cubic in pupil
Astigmatism	$C_{ast1}\,\alpha^2 y^2$	Quadratic in both field and pupil
Field curvature	$C_{FC}\,\alpha^2(x^2+y^2)$	Quadratic in field and pupil
Distortion	$C_{dist}\,\alpha^3 y$	Cubic in field, linear in pupil

It is convenient to expand W as a polynomial in terms of $x^n y^m (h/z)^k$. Restricting the series to terms with $n+m+k \le 4$ yields a polynomial of 4th order in $x, y, h/z$. The truncated expansion is

$$W \approx h^2\frac{1}{2z} + h\frac{y}{z} + \frac{R+z}{2Rz}(x^2+y^2) + \frac{z^3+R^3}{8(R^3z^3)}(x^2+y^2)^2 - \frac{h}{2z^3}y(x^2+y^2) + \frac{h^2}{4z^3}(x^2+y^2) + \frac{h^2}{2z^3}y^2 - \frac{h^3}{2z^3}y. \tag{4.3}$$

Equation 4.3 describes the fourth-order wavefront aberrations of a single spherical surface and highlights the dependence of W on both the pupil coordinates (x, y) and the field angle

$$\alpha = h/z. \tag{4.4}$$

Table 4.1 summarizes these relations for a general optical system. Third-order transverse aberrations are obtained by differentiating the wavefront aberrations with respect to (x, y) and scaling by the system focal length. The five primary aberrations are illustrated schematically in Fig. 4.2.

The coefficients C in Table 4.1 depend on the specific system configuration and can be evaluated by tracing paraxial rays and calculating the so-called *Seidel sums*. A minimum of one ray suffices, but it is more practical to trace both the chief ray (from the field edge through the pupil center) and the marginal ray (from the object axis through the pupil edge). While Seidel analysis requires little computation compared to exact ray tracing, the resulting expressions are algebraically cumbersome. Moreover, eliminating third-order aberrations is necessary but not sufficient for good imaging. At large numerical apertures or wide fields of view, higher-order aberrations become significant, necessitating full ray-tracing and system optimization.

Nonetheless, the scaling relations in Table 4.1 allow quick engineering estimates. For example, if a lens exhibits certain amounts of astigmatism and coma at a field angle α, increasing α will cause astigmatism to grow approximately quadratically, while coma will grow linearly. Similar estimates hold for other aberrations.

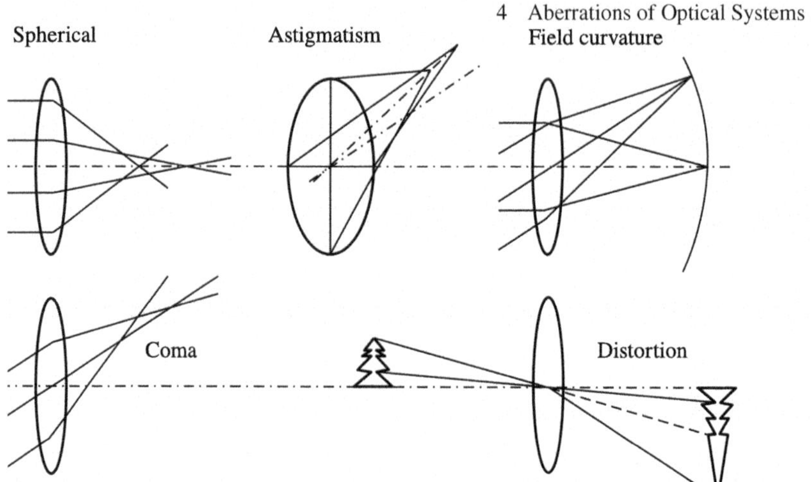

Fig. 4.2 Schematic ray diagrams for the five primary aberrations

Several general properties also follow from the analysis of Seidel sums:

- Spherical aberration is independent of pupil position and field angle.
- If spherical aberration is corrected, coma does not vary with pupil position.
- If spherical aberration and coma are corrected, astigmatism is independent of pupil position.
- If spherical aberration, coma, and astigmatism are corrected, distortion becomes independent of pupil position.
- The field curvature radius R_F is independent of stop position and can be estimated via the Petzval sum:

$$\frac{1}{R_F} = -n_k \sum_{i}^{k} \frac{n_i - n_{i-1}}{r_i n_i n_{i-1}}, \tag{4.5}$$

  where r_i is the radius of curvature of the ith surface separating media of indices n_{i-1} and n_i. Positive surfaces contribute to negative field curvature (image edges shifted toward the object), while negative surfaces contribute to positive curvature. Equation 4.5 is valid only if astigmatism is fully corrected.
- Aberrations depending on odd powers of the field angle α (coma, distortion) are compensated in symmetrical systems.

In general, the strong dependence of aberrations on system parameters makes optical design an intricate optimization problem. Only very simple systems can be designed analytically. A well-known example is the *Cooke's triplet*, consisting of three lenses, which can be analytically corrected for all third-order and chromatic aberrations. Analytical methods of design and aberration balancing were extensively developed in the 20th century. Modern highly corrected lenses rely on computer optimization, yet the preliminary analysis of third-order aberrations remains fundamental to the conceptual design of optical systems.

4.2 Zernike Polynomials

The aberration functions listed in Table 4.1 are not mutually orthogonal. For practical analysis it is convenient to employ aberration functions that are orthonormal over the pupil. Since the pupil of most optical systems is circular, these functions should be orthonormal within a unit circle. A set of functions W_i is orthonormal in a circle if

$$\int_S W_i W_j \, ds = 0 \quad \Big|_{i \neq j},$$
$$\int_S W_i W_j \, ds = \pi \quad \Big|_{i = j}, \tag{4.6}$$

where S denotes the unit circle. If the total aberration across the pupil is described by the function V, then V can be decomposed into a sum of orthogonal polynomials:

$$V = \sum_i a_i W_i, \quad \text{where}$$
$$a_i = \frac{1}{\pi} \int_S W_i V \, ds. \tag{4.7}$$

Zernike polynomials form such an orthonormal basis in the unit circle. Low-order Zernike polynomials closely resemble the classical third-order aberrations, which makes them especially suitable for representing optical wavefronts (Table 4.2 and Fig. 4.3).

Unlike the third-order Seidel aberrations, Zernike polynomials are defined only over the pupil and do not explicitly include the object coordinate given in Eq. 4.4.

Table 4.2 Low-order Zernike polynomials. $\rho = x^2 + y^2, \quad \varphi = \arctan(y/x)$

Indices	Expression	Aberration
0 0	1	Piston
1 1	$2\rho \cos \varphi$	Tilt (distortion)
1 –1	$2\rho \sin \varphi$	Tilt
2 0	$\sqrt{3}(2\rho^2 - 1)$	Defocus (field curvature)
2 2	$\sqrt{6}\rho^2 \cos(2\varphi)$	Astigmatism
2 –2	$\sqrt{6}\rho^2 \sin(2\varphi)$	Astigmatism
3 1	$\sqrt{8}(3\rho^3 - 2\rho) \cos(\varphi)$	Coma
3 -1	$\sqrt{8}(3\rho^3 - 2\rho) \sin(\varphi)$	Coma
3 3	$\sqrt{8}\rho^3 \cos(3\varphi)$	Trefoil
3 –3	$\sqrt{8}\rho^3 \sin(3\varphi)$	Trefoil
4 0	$\sqrt{5}(6\rho^4 - 6\rho^2 + 1)$	Spherical

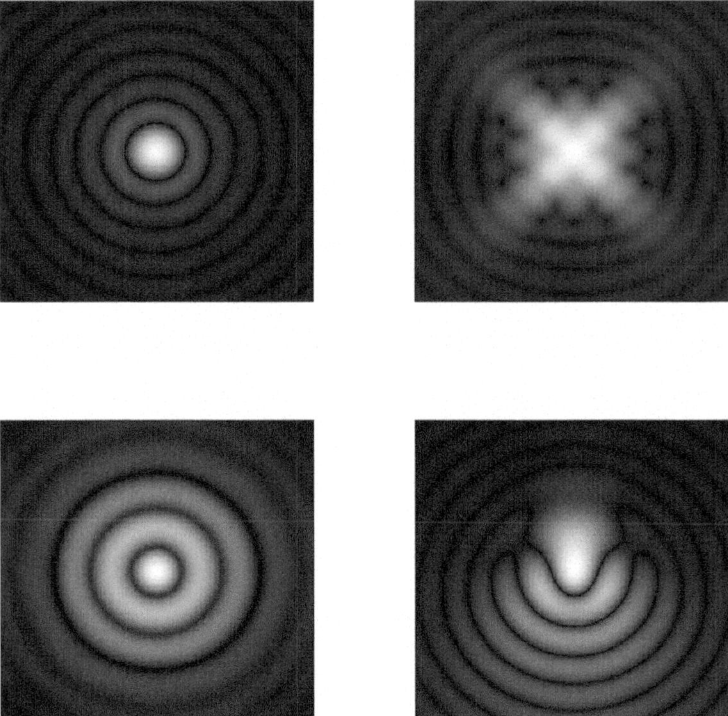

Fig. 4.3 Far-field image of a monochromatic point source with no aberration (top left), with half-wave of astigmatism (top right), half-wave of spherical aberration (bottom left), and half-wave of coma (bottom right). The intensity of diffracted fringes is enhanced using gamma correction with $\gamma = 3$: $I_{out} = \left(\frac{I_{in}}{I_{in\,max}}\right)^{1/\gamma}$. All images are normalized to their maximum intensity

Seidel polynomials describe the wide-field behavior of an optical system, whereas Zernike polynomials characterize the aberration at a single point in the field.

As a result, the same Zernike polynomial may correspond to different aberrations. For example, the defocus term also represents field curvature: in pure defocus, the aberration amplitude is constant across all field positions, while in the case of field curvature the defocus amplitude varies with the object angle α.

4.3 Spherical Aberration of a Thin Lens

The longitudinal spherical aberration (focus error) for a thin lens of refractive index n, formed by two spherical surfaces with curvatures $\rho_1 = R_1^{-1}$ and $\rho_2 = R_2^{-1}$, is

$$\Delta F = \frac{y^2}{2n(n-1)(\rho_1-\rho_2)}\left[(n+2)\rho_1^2 - n(2n+1)\rho_1(\rho_1-\rho_2) + n^3(\rho_1-\rho_2)^2\right], \tag{4.8}$$

where y is the ray height in the lens plane.

For an object at infinity, the condition for minimum spherical aberration in a lens with focal length F_0 is obtained when

$$R_1 = 2F_0\frac{(n+2)(n-1)}{n(2n+1)},$$
$$R_2 = 2F_0\frac{(n+2)(n-1)}{(2n^2-n-4)}. \tag{4.9}$$

For $n \approx 1.5$ these relations give $R_1 \approx 0.583F_0$ and $R_2 \approx -3.5F_0$. Thus, to minimize spherical aberration, the more strongly curved surface should face the direction of infinity, whether on the object side or on the image side.

It follows from Eq. 4.8 that the minimal spherical aberration of a system composed of two identical lenses with combined focal length F is smaller than that of a single lens of the same focal length.

The longitudinal spherical aberration of a thin parallel plate of thickness T and refractive index n is

$$\Delta = \frac{n^2-1}{n^3}TA^2, \tag{4.10}$$

where A is the numerical aperture of the incident cone of rays.

4.4 Aplanatic Points and Sine Condition

Aplanatic points are pairs of conjugate points that are imaged free from spherical aberration and coma. Two on-axis conjugate points O and O' are aplanatic if the optical system satisfies Abbe's sine condition, illustrated in Fig. 4.4:

$$n \sin\varphi = n' \sin\varphi'. \tag{4.11}$$

The sine condition is of particular importance in the design of optical systems with large numerical apertures ($A > 0.5$). In this regime the third-order aberration theory loses accuracy, and the sine condition provides a simple criterion to ensure that spherical aberration and coma are absent near the optical axis.

A spherical mirror possesses two pairs of aplanatic points:

- The center of curvature is imaged into itself.
- The mirror apex is imaged into itself.

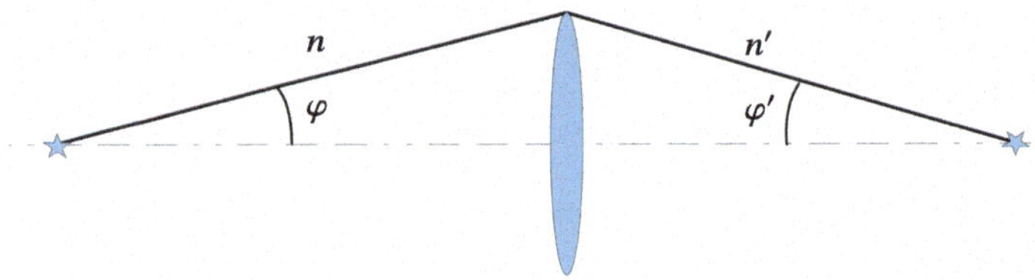

Fig. 4.4 Illustration of Abbe's sine condition $n \sin \varphi = n' \sin \varphi'$. For non-paraxial, aplanatic imaging of an on-axis point, the condition must be satisfied for all ray angles φ

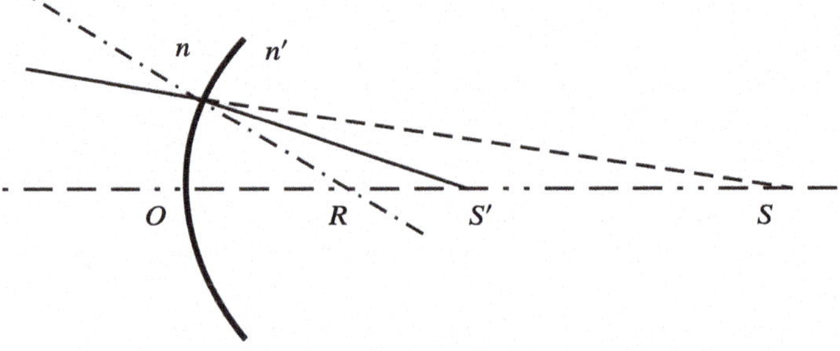

Fig. 4.5 Aplanatic points of a spherical refracting surface. The apex O and the center of curvature R are self-conjugate. Points S and S' are conjugate if $|OS| = R(1 + n'/n)$ and $|OS'| = R(1 + n/n')$

A spherical refracting surface has three pairs of aplanatic points:

1. A spherical wavefront converging to the center of curvature experiences no spherical aberration, since all rays strike the surface normally. The point at the center of curvature is therefore aplanatic.
2. For a spherical interface of radius R between media of indices n and n', two conjugate aplanatic points exist at $S = R(1 + n'/n)$ and $S' = R(1 + n/n')$, as shown in Fig. 4.5.
3. The apex of the surface is imaged into itself, forming another self-conjugate aplanatic point.

A variety of aplanatic lens configurations can be realized by combining surfaces of the first and second types. The second type of aplanatic design is widely applied in immersion microscopy, where it reduces the numerical aperture of the light cone crossing the glass-air interface, in a manner analogous to the imaging of point S' in Fig. 4.5.

4.5 Conic Sections in Optics

An ellipse is a conic section characterized by two foci. For any point on its contour, let l_1 be the distance to the first focus and l_2 the distance to the second. Then, $l_1 + l_2 = \text{const}$ for all contour points. This property has a direct optical interpretation:

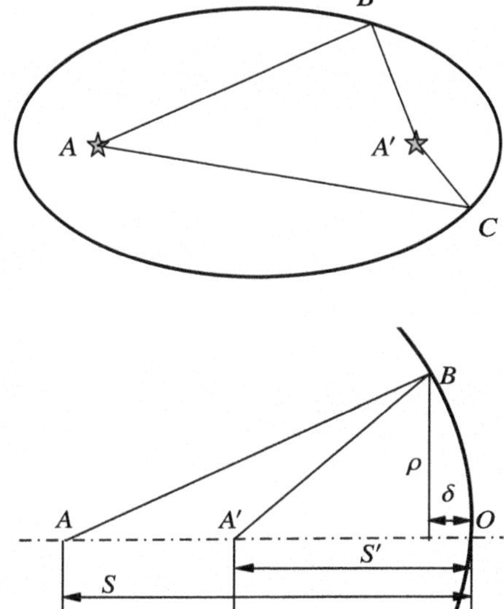

Fig. 4.6 The path length $|ABA'|$ equals $|ACA'|$, where B and C are arbitrary points on an ellipse. The focal points A and A' are conjugate

Fig. 4.7 Reflection from an aspheric mirror

rays emitted from one focus and reflected by the inner surface of an ellipse converge exactly at the second focus, since their optical paths are equal. Thus, an elliptic mirror provides perfect imaging between its two focal points, as illustrated in Fig. 4.6.

Using this property, one can derive the shape of a mirror that provides aberration-free imaging between two arbitrary points S and S'. To satisfy the equal optical path condition (1.1), the relation

$$\sqrt{(S-\delta)^2+\rho^2}+\sqrt{(S'-\delta)^2+\rho^2}=S+S' \tag{4.12}$$

must hold for any point B, as shown in Fig. 4.7. The solution is

$$\delta=\mp\frac{(S'+S)\sqrt{S^2\,S'^2-\rho^2 SS'}\;\mp SS'^2\mp S^2\,S'}{2SS'}, \tag{4.13}$$

where the minus sign corresponds to real imaging ($SS'>0$) and the plus sign to virtual imaging.

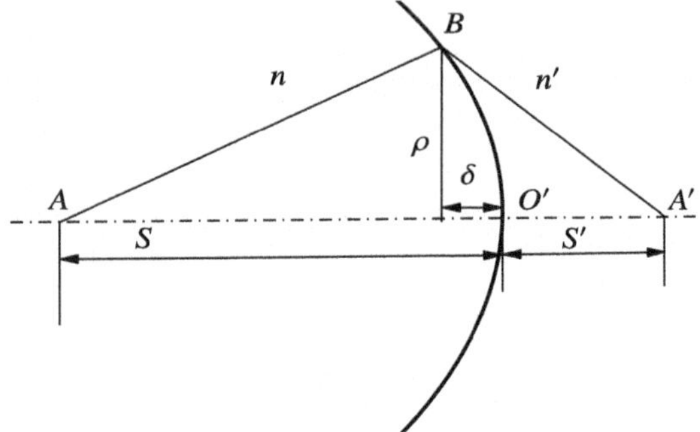

Fig. 4.8 Refraction on a single aspheric surface

In optical engineering, Eq. 4.13 is often expressed in standard form:

$$\delta = \frac{\rho^2}{R\left(\sqrt{1 - \frac{\rho^2(e^2+1)}{R^2}} + 1\right)},$$
$$e^2 = -\left[\frac{S - S'}{S + S'}\right]^2,$$
$$R = 2\frac{SS'}{S + S'}, \tag{4.14}$$

where R is the paraxial radius of curvature and e^2 is the conic parameter, related to the eccentricity, that defines the asphermization of the surface.

Refraction at an aspheric surface can be analyzed similarly. The optical path $|ABA'|$ must equal $|AOA'|$, as illustrated in Fig. 4.8:

$$n\sqrt{(S - \delta)^2 + \rho^2} + n'\sqrt{(S' + \delta)^2 + \rho^2} = nS + n'S'. \tag{4.15}$$

This equation admits a conic-section solution only for $S = \pm\infty$, yielding

$$n'\sqrt{(S' + \delta)^2 + \rho^2} = n\delta + n'S', \tag{4.16}$$

with solution

$$\delta = \pm\frac{n'\sqrt{(n'^2 - 2nn' + n^2)S'^2 - \rho^2(n'^2 - n^2)} \pm (nn' - n'^2)S'}{n'^2 - n^2}, \tag{4.17}$$

where the plus sign applies if $(n - n')S' < 0$.

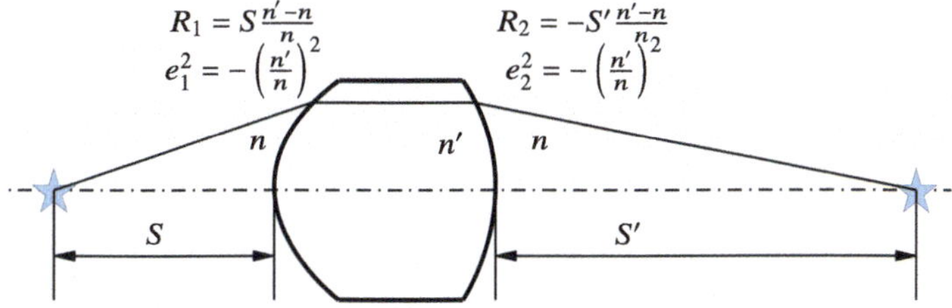

Fig. 4.9 Design of a finite-conjugate lens free of spherical aberration

In standard form this becomes

$$\delta = \frac{\rho^2}{R\left(\sqrt{1 - \frac{\rho^2(e^2+1)}{R^2}} + 1\right)},$$
$$e^2 = -\left(\frac{n}{n'}\right)^2,$$
$$R = S'\frac{n - n'}{n'}. \tag{4.18}$$

To eliminate spherical aberration of a single refractive surface with finite conjugates ($|SS'| < \infty$), the refracting surface must be described by a higher-order aspheric polynomial rather than a simple conic. However, aberration-free finite-conjugate imaging can be achieved with a lens consisting of two aspheric surfaces described by conic sections, as shown in Fig. 4.9. In this case the beam is collimated inside the lens, so each surface satisfies the infinite-conjugate condition.

4.6 Chromatic Aberrations

The refractive index of all optical materials depends on wavelength. This dependence, denoted $n(\lambda)$, is called *dispersion*. Dispersion is particularly strong in the vicinity of absorption lines. Since most optical glasses absorb at shorter wavelengths, the refractive index increases as λ decreases. The function $n(\lambda)$ is highly nonlinear and must be determined experimentally. Several approximations are widely used in optical engineering, the most common being:

the Hartmann dispersion formula,

$$n = A + \frac{B}{(\lambda - C)^2}, \tag{4.19}$$

the Sellmeier formula,

$$n^2 = 1 + \sum_{i=1}^{3} \frac{B_i \lambda^2}{\lambda^2 - C_i}, \tag{4.20}$$

and the Schott approximation,

$$n^2 = A_0 + A_1\lambda^2 + \frac{A_2}{\lambda^2} + \frac{A_3}{\lambda^4} + \frac{A_4}{\lambda^6} + \frac{A_5}{\lambda^8}. \tag{4.21}$$

A single-valued measure of glass dispersion is given by the *Abbe number* ν, defined as

$$\nu = \frac{n_D - 1}{n_F - n_C}, \tag{4.22}$$

where n_D, n_F, n_C are the refractive indices measured at wavelengths $\lambda_D = 589.3$ nm, $\lambda_F = 486.1$ nm, and $\lambda_C = 656.3$ nm. Different reference lines are sometimes used to compute ν, so care must be taken when applying tabulated values in calculations.

Flint glasses, with high refractive index $n \sim 1.5 \ldots 2$, exhibit strong dispersion and therefore have low Abbe numbers $\nu \sim 20 \ldots 40$.

Crown glasses, with lower refractive index $n \sim 1.4 \ldots 1.7$, show weaker dispersion and correspondingly higher Abbe numbers $\nu \sim 50 \ldots 70$. Materials with very low dispersion, such as fluorite, may reach Abbe numbers near 90.

Optical glasses are also classified by a six-digit glass code (also called glass number or MIL code), standardized in MIL-G-174. The first three digits correspond to $n_D - 1$, rounded to three digits, while the last three digits correspond to the Abbe number ν. For example, BK7 with $n_D = 1.5168$ and $\nu = 64.17$ has the MIL code 517642. A linear approximation of $n(\lambda)$ near the D line is

$$n(\lambda) = n_D + \frac{n_D - 1}{170.2\nu}(589.3 - \lambda), \tag{4.23}$$

where λ is in nm.

This approximation provides a quick estimate of dispersion behavior in the visible range, with errors below $5 \cdot 10^{-4}$ over $\lambda_F \ldots \lambda_C$. While insufficient for precision design, it is useful for preliminary analysis. More accurate results can be obtained using the first three terms of the Schott formula (4.21):

$$n^2 = A + B\lambda^2 + \frac{C}{\lambda^2}, \tag{4.24}$$

where coefficients A, B, C are determined from refractive indices at three spectral lines. Although the closed-form expressions for these coefficients are cumbersome, they can be readily evaluated by computer algebra software.

Since all transparent materials are dispersive, the focal length of a lens depends on wavelength. For a thin lens with focal length F defined as

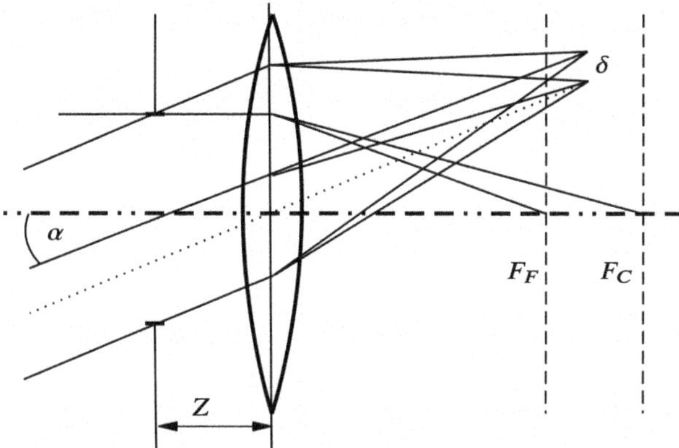

Fig. 4.10 Longitudinal chromatic aberration: a collimated polychromatic beam along the optical axis produces two foci, F_F (blue) and F_C (red). Off-axis, magnification error appears: a beam at angle α is focused differently for blue and red, producing a transverse shift δ in addition to the confusion circle from longitudinal aberration

$$\frac{1}{F} = \left(\frac{1}{R_1} - \frac{1}{R_2}\right)(n-1), \quad \text{or} \quad F = \frac{R_1 R_2}{(n-1)(R_1 - R_2)}, \tag{4.25}$$

the focal positions for wavelengths $\lambda_F = 486.1$ nm and $\lambda_C = 656.3$ nm differ by

$$\delta = \frac{dF}{dn} \cdot (n_C - n_F) \approx \frac{F}{\nu}, \tag{4.26}$$

Thus as shown in Fig. 4.10, the chromatic focal error of a single lens is approximately ν times smaller than its focal length. Expressed in diopters, the chromatic error is

$$\Phi_C = \frac{1}{F\nu} = \frac{\Phi}{\nu}, \quad \text{where} \quad \Phi = \frac{1}{F}. \tag{4.27}$$

Chromatic focus error, see Fig. 4.10, also gives rise to transverse chromatic aberration. Point images formed by different wavelengths produce distinct confusion circles in the focal plane. The longitudinal chromatic aberration of a thin element is

$$dS' = \frac{S'^2}{\nu F'}, \tag{4.28}$$

and the radius of the confusion circle ρ is

$$\rho \approx A' dS' = A' \frac{S'^2}{\nu F'}, \tag{4.29}$$

where A' is the numerical aperture in image space. This estimate assumes one wavelength is perfectly focused while the other forms the maximum blur circle. At the best-focus position, when blur circles of different wavelengths are balanced, the effective radius is about half of that defined by Eq. 4.29.

Achromatization can be achieved for certain thick-lens configurations. Differentiating the lensmaker's formula (3.18) with respect to n_1 and setting the result to zero yields the condition

$$\frac{R_1 - R_2}{D} = 1 - \frac{1}{n_1^2}, \tag{4.30}$$

defining a *negative achromatic meniscus*. By adjusting R_1 and R_2 while maintaining Eq. 4.30, the lens shape can be tuned to control spherical aberration as well.

At wide field angles, wavelength-dependent magnification differences produce *chromatic magnification error*, observed as color fringing in off-axis image regions. The amount of fringing is proportional to the field angle α. The shift δ between red and blue images of an off-axis point formed by a thin component is estimated as

$$\frac{\delta}{H'} = -\frac{1}{F\nu(S^{-1} - Z^{-1})}, \tag{4.31}$$

where Z is the distance from the entrance pupil to the principal plane, S is the object distance, and H' is the image height ($Z = S$).

In summary, chromatic aberration arises from glass dispersion and can be observed as a shift of focal positions (longitudinal), and as color fringing across the field. Chromatism is corrected by using combinations of crown and flint glasses in achromatic lenses. These corrections are essential for achieving high image quality across the visible spectrum. Mirror systems are free from chromatic aberrations.

4.7 Achromatic Doublet

Chromatic aberration can be compensated by combining two lenses so that the chromatic error of the first lens (with focal length F_1 and Abbe number ν_1) is canceled by that of the second lens (with focal length F_2 and Abbe number ν_2). The total focal length F of the system is:

$$F = \frac{F_1 F_2}{F_1 + F_2}. \tag{4.32}$$

To minimize chromatic aberration, we require

$$\frac{1}{F_1\nu_1} + \frac{1}{F_2\nu_2} = 0 \quad \text{or} \quad F_2 = -F_1\frac{\nu_1}{\nu_2}. \tag{4.33}$$

Solving Eqs. 4.32 and 4.33 together yields the component focal lengths:

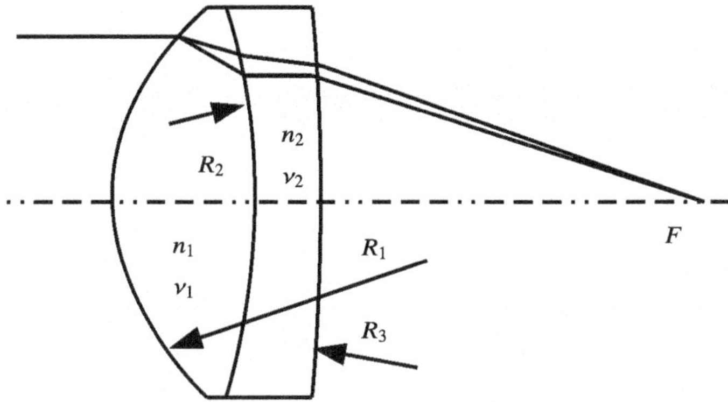

Fig. 4.11 Achromatic doublet

$$F_1 = F\frac{\nu_1 - \nu_2}{\nu_1}, \quad F_2 = -F\frac{\nu_1 - \nu_2}{\nu_2} \quad \text{or}$$
$$\Phi_1 = \frac{1}{F_1} = \Phi\frac{\nu_1}{\nu_1 - \nu_2}, \quad \Phi_2 = \frac{1}{F_2} = -\Phi\frac{\nu_2}{\nu_1 - \nu_2}. \tag{4.34}$$

Equations 4.34 can be applied to the design of a doublet lens (Fig. 4.11) with focal length F, made of two glasses (n_1, ν_1) and (n_2, ν_2), corrected for chromatic aberration.

For objects at infinity, spherical aberration is minimized by a lens shape with a strongly convex front surface and an almost flat back surface. We assume the back surface is flat ($R_3 = \infty$), which is a good approximation when the aperture is small. According to Eq. 4.34, the component with the higher Abbe number (crown glass) must be positive, while the one with the lower Abbe number (flint glass) must be negative. Crown glasses are more scratch-resistant and chemically stable; therefore, in systems exposed to harsh conditions (rain, dust, etc.), the first convex element is typically crown, followed by a negative flint element.

Assuming $R_3 = \infty$ and combining Eqs. 4.25 with 4.34, we obtain the radii of curvature:

$$R_2 = \frac{F(\nu_2 - \nu_1)(n_2 - 1)}{\nu_2},$$
$$R_1 = \frac{(\nu_2 - \nu_1)(n_2 - 1)(n_1 - 1)F}{(n_1 - 1)\nu_2 - (n_2 - 1)\nu_1}. \tag{4.35}$$

A special case is $R_1 = -R_2$, giving the so-called "Littrow doublet", which is simple to manufacture since all surfaces have the same curvature. The glasses for a Littrow doublet satisfy:

$$n_1 = 1 + (n_2 - 1)\frac{\nu_1}{2\nu_2}.$$

Table 4.3 Achromatic doublet formed by a BK7–F2 glass pair. $F = 50,\ n_2 = 1.62,\ \nu_2 = 36.36,\ n_1 = 1.5168, \nu_1 = 64.17$

Parameter	Estimate from Eq. 4.35	Zemax optimization
R1	21.22	22.08
d1	–	2
R2	–23.7	–23.5
d2	–	1
R3	∞	∞

No real glass pair fulfills this exactly, but practical approximations exist, e.g. SK3 (607597) with KZFS8 (721347), or N-PSK57 (592684) with N-LAF32 (794455) from the Schott catalog.

For the more general case $R_3 \neq \infty$, additional (though more complex) expressions can be derived, with R_3 providing an extra degree of freedom for correcting spherical aberration (see Eq. 4.8).

Table 4.3 shows that Eq. 4.35 provides results in close agreement with direct ray-tracing. Both designs are achromatic and diffraction-limited on-axis for $F\# \geq 7$.

Because glass dispersion is nonlinear, a doublet can only be made achromatic at two wavelengths. Residual chromatic error at other wavelengths is called the *secondary spectrum*. Unlike monochromatic aberrations, chromatic aberrations are extremely difficult to remove completely by adding more lenses, because most glasses have very similar dispersion curves. This similarity limits the number of independent correction functions available, making full achromatization challenging.

4.8 Performance Characterization

The performance of a corrected optical system is evaluated at different positions within the field of view and across the specified spectral range. As an illustration, we consider the case of a triplet lens shown in Fig. 4.12.

To characterize wavefront aberrations, the nearest reference sphere is subtracted from the actual wavefront. The resulting wavefront error is then computed over the system pupil. Since the wavefront error depends on both field position and wavelength, system optimization aims to minimize it across all relevant fields and spectral bands. An example of a wavefront map, dominated by spherical aberration, is shown in Fig. 4.13.

The *Strehl parameter* S, defined as the normalized loss of far-field axial intensity due to aberrations, is related to the root mean square (*rms*) wavefront deviation σ (in radians) from the best-fit reference sphere as

$$S \simeq e^{-\sigma^2}. \tag{4.36}$$

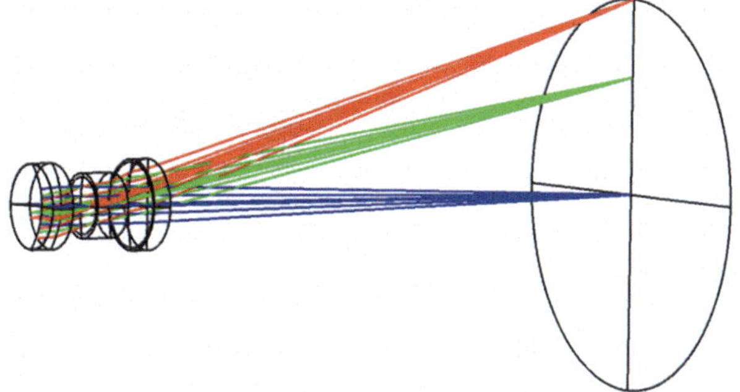

Fig. 4.12 Layout of a triplet lens

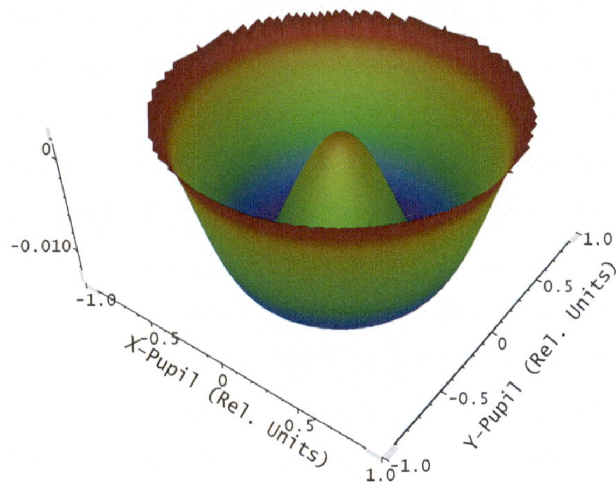

Fig. 4.13 Wavefront map for the on-axis field position

Although S is defined for any value of σ, the exponential approximation is valid only for small aberrations with $\sigma \ll 1$.

Point spread functions (PSF) and ray scatter diagrams provide further insight into image plane performance. A diffraction-limited system focuses all geometrically traced rays within the Airy disk for all field positions and wavelengths. Since this condition is rarely achieved in practice, the deviation between the actual and diffraction-limited PSF is an indicator of image quality. An example of spot diagrams for a triplet lens is shown in Fig. 4.14.

Mathematically, the PSF $I(\rho)$ is the squared modulus of the Fourier transform of the complex pupil function $U(\xi)$:

$$I(\rho) = | F(U(\xi))|^2, \tag{4.37}$$

where F denotes the Fourier transform.

The Fourier transform of the PSF defines the *Optical Transfer Function (OTF)* $O(\xi)$:

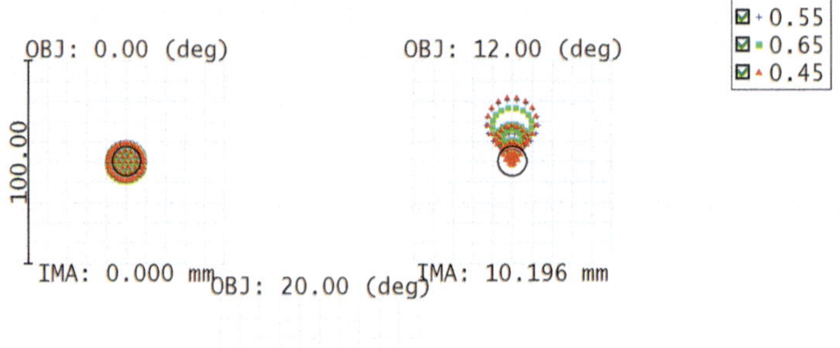

Fig. 4.14 Spot diagram showing scatter plots for different fields and wavelengths. The Airy disk is indicated by the black circle. The diagram clearly demonstrates that the lens is not diffraction limited

$$O(\xi) = F(I(\rho)) = \int_S U(w)\, U(w+\xi)\, dw, \tag{4.38}$$

whose modulus is the *Modulation Transfer Function (MTF)*. The MTF describes how image contrast $(I_{max} - I_{min})/I_{max}$ varies with spatial frequency ξ.

For an incoherent diffraction-limited optical system, the MTF acts as a spatial frequency filter with cutoff

$$\xi_c = \frac{1}{\lambda F\#}. \tag{4.39}$$

Expressed in terms of normalized spatial frequency $\xi = \nu\lambda F\#$, the MTF of an axially symmetric diffraction-limited system is

$$M_D(\xi) = \frac{2}{\pi}\left[\arccos(\xi) - \xi\sqrt{1-\xi^2}\right]. \tag{4.40}$$

For a system with a square aperture, the MTF reduces to

$$M_D(\xi) = 1 - \xi. \tag{4.41}$$

In real systems the MTF at any frequency ξ is always lower than the diffraction limit $M_D(\xi)$. In multi-element systems the PSF, and hence the MTF, varies with both field position and wavelength. A common way to specify performance is by plotting the MTF at fixed spatial frequencies (e.g., $\nu = 20, 40, 60$ line pairs/mm) across the field. Figures 4.15 and 4.16 illustrate typical MTF behavior of an aberrated system.

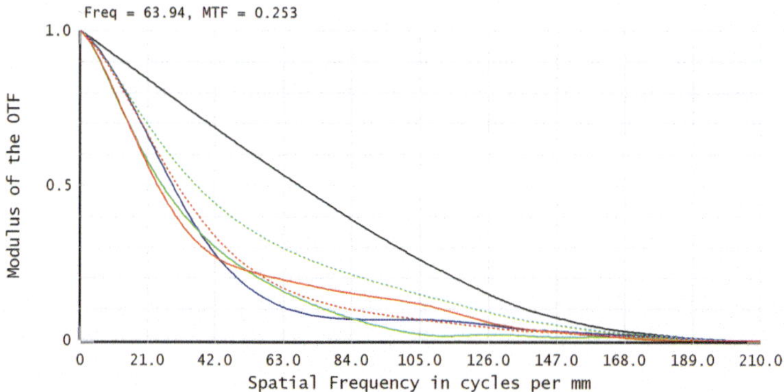

Fig. 4.15 MTF of the triplet system, calculated using a Fast Fourier Transform (FFT). Colored curves correspond to different field positions, while the black curve shows the diffraction limit

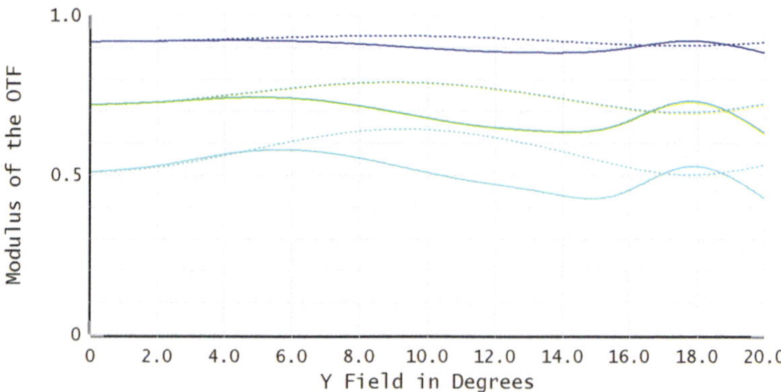

Fig. 4.16 MTF versus field position for spatial frequencies of 10, 20, and 30 line pairs/mm

For laser beams, performance is characterized not by the MTF but by the M^2 *parameter* and by the "power-in-the-bucket" metric, which describes the fraction of total beam power captured within a predefined circular aperture of specified angular or linear radius.

4.9 Conclusion

Aberration can be defined as a deviation of the actual wavefront W_a from the ideal spherical wavefront W_i, expressed through the aberration function $W = W_a - W_i$. Expansion of W into low-order polynomials yields the third-order aberration terms corresponding to five primary monochromatic aberrations: spherical, coma, astigmatism, field curvature, and distortion. Their scaling with pupil coordinates and field angle provide analytical estimates of the performance of optical system.

Zernike polynomials form an orthogonal basis over the circular pupil and are widely used to quantify aberrations. Low-order Zernike terms correspond directly to classical aberrations.

A single spherical refracting surface possesses two conjugate aplanatic points where coma and spherical aberration vanish. General conic sections and higher-order aspheres allow finite-conjugate imaging without spherical aberration.

Chromatic aberrations result from the refractive index dispersion $n(\lambda)$. The Abbe number ν serves as a practical measure of dispersion. Longitudinal and transverse chromatic errors scale as $1/\nu$ and can be reduced by combining crown and flint glasses. The achromatic doublet cancels primary chromatic error for two wavelengths, leaving only the chromatic aberration due to the secondary spectrum.

Image quality is characterized by the point spread function (PSF), the optical transfer function (OTF), and the modulation transfer function (MTF). These functions link the aberrations to the spatial frequency response and image contrast. The Strehl ratio provides a compact performance metric, approximated by $S \simeq e^{-\sigma^2}$ for *rms* wavefront error $\sigma \ll 1$.

4.10 Problems

4.1 The optometrist has offered you a choice of two different spectacles: the first one is "cheap", as it has thicker lenses made of material with low refraction and high Abbe number. The other one is "modern" as it has thin lenses made of material with high refraction and lower Abbe number. Compare these glasses with respect to (1) spherical aberration and (2) chromatic aberrations.

Solution (4.1). Modern glasses excel in minimizing the spherical aberration but suffer from higher chromatism. Cheap glasses reduce chromatic aberration but have larger spherical aberration.

4.2 Business producing ophthalmic instruments for measurement of the refractive error claims the measurement precision of $\delta P = 0.01$ diopter. Assuming pupil diameter of 7 mm, calculate the corresponding amplitude of the wavefront aberration.

Solution (4.2). In the paraxial approximation the converging wavefront is described by

$$W = \rho^2/2F = \rho^2/2 \cdot P,$$

where ρ is the radial coordinate, and $P = 1/F$ is the focusing power in diopter. Then,

$$\Delta W = \frac{dW}{dP}\Delta P = \rho^2/2 \cdot \delta P.$$

Substituting $\rho_{max} = 3.5$ mm, and $\delta P = 0.01$, we obtain $\Delta W = 61.25$ nm.

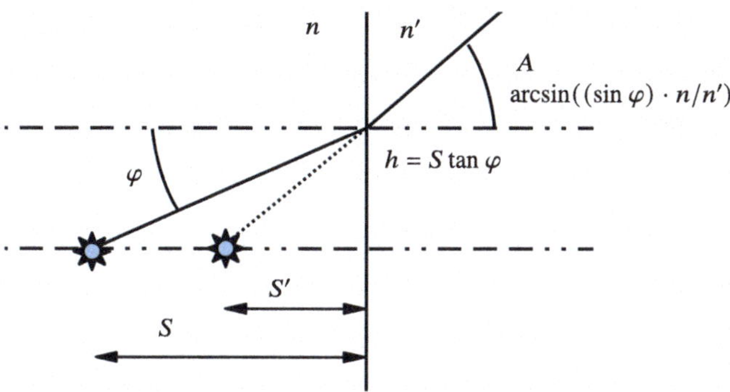

Fig. 4.17 To Problems 4.4 and 4.5

4.3 The photographic lens set to $F\# = 4$ has equal peak-to-valley (PV) wavefront amplitudes of coma, astigmatism and spherical aberration. Estimate the change of the amplitude of these aberrations if the system diaphragm is set to (A) $F\# = 2$, and (B) if the image linear size (FOV) is reduced 2 times at $F\# = 4$.

Solution (4.3). According to Table 4.1, changing the $F\#$ from 4 to 2 will increase the PV amplitude of coma by 8 times, increase the amplitude of astigmatism by 4 times, and increase the amplitude of spherical aberration by 16 times.

If the field of view is reduced by 2 times, the PV amplitude of coma will become 2 times smaller, the PV amplitude of astigmatism will become 4 times smaller, and the amplitude of spherical aberration will not change at all.

4.4 Show that a flat interface between refractive indices n and n' introduces aberration into a spherical wave.

Solution (4.4). Ray originating from a point at a distance S from the interface in the medium with refraction index n_1 and coming to the interface with angle φ will be seen from the second medium as if it originates from a point at a distance

$$S' = S \frac{\tan \varphi}{\tan(\arcsin((\sin \varphi) \cdot n/n'))}$$

from the interface as shown in Fig. 4.17. Since S' depends on φ, the homocentricity of the beam is not preserved.

4.5 Estimate the amount of longitudinal ray aberration introduced in a microscopy sample by a cover glass with thickness $S = 0.17$ mm and $n = 1.5$, if the sample is observed using objective with numerical aperture A.

Solution (4.5). Applying the Snell's law to the rays originating from the point at depth S inside glass, the difference δS of the sample position given by the paraxial and the marginal rays is

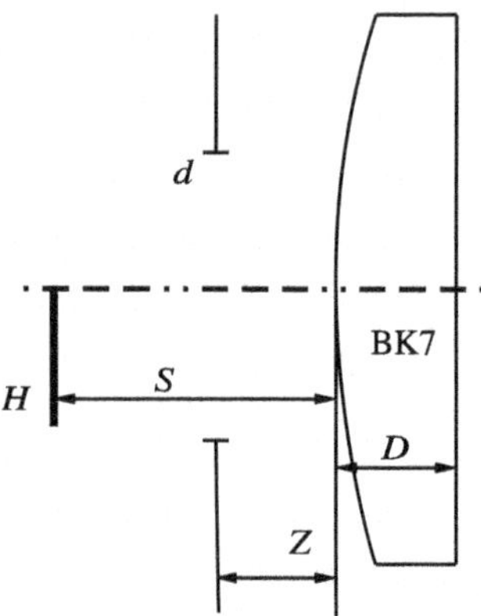

Fig. 4.18 To Problem 4.6

$$\delta S' = S[\frac{1}{n} - \frac{\tan(\arcsin A)}{\tan(\arcsin((An))}].$$

For A = 0.5, n = 1.5 and S = 0.17 mm, $\delta S' = 26\ \mu\text{m}$. For high quality imaging, the absolute value of $\delta S'$ should be smaller than the DOF of the objective. Therefore, for high values of A, the objective should be corrected for the spherical aberration of the cover glass. To find the maximum value of A, we equate the longitudinal focal spread to the depth of field:

$$|\delta S'| < \frac{n\lambda}{A^2}.$$

The approximate solution can be obtained by using the second order Taylor series for $\delta S'$: $\delta S' \approx A^2 S(n^2 - 1)/(2n)$:

$$A \lesssim \left[\frac{2\lambda n^2}{S(n^2 - 1)}\right]^{1/4}.$$

For $\lambda = 0.55\ \mu\text{m}$, $S = 170\ \mu\text{m}$, and $n = 1.5$, we have $A \lesssim 0.32$.

Similar result can be obtained equating the longitudinal spherical aberration given by Eq. 4.10 to the depth of field.

4.6 Object with height H is imaged by a plano-convex lens fabricated from BK7 glass with MIL code 517642. Aperture stop with diameter of $d = 10$ mm is positioned in front of the lens at a distance of $Z = 10$ mm, as shown in Fig. 4.18. The object size is $H = 10$ mm, the object distance to the front surface of the lens is $S = 100$ mm, the lens thickness is $D = 5$mm, and the radius of curvature of the front surface $R = 25$ mm. Calculate the image position S'_b referenced to the back surface of the lens, estimate the diameter of confusion circle due to chromatic aberration, and maximum fringing due to chromatic shift.

Solution (4.6). For a plano-convex lens the focal length $F = R/(n-1) = 25/(0.517) = 48.35$. The distance from the back principal plane to the image is given by

$$S' = \frac{FS}{S-F} = \frac{48.35100}{100-48.35} = 93.61.$$

The back focal plane is positined at a distance $d = -D/n = -5/1.517$ relative to the back surface of the lens. Finally, the distance from the back surface to the image is $S' - D/n = 90.314$ mm.

The image size $H' = HS'/S = 9.03$ mm. The front principal plane is co-incident with the top of the lens surface. Using the Lagrange invariant, we state: $HA = H'A'$, where A and A' are the object and image space numerical apertures, and using $H' = HS'/S$ we obtain $A' = HAS/HS' = AS/S' = 0.5d/(S-Z)S/S' = 0.0593$. According to Eq. 4.29, he radius of the chromatic circle of confusion in the optimum focus position is

$$r \approx 0.5 \cdot A'S'^2/\nu/F \approx 0.5 \cdot 0.0593 \cdot 93.61^2/64.2/48.35 = 0.078 \text{ mm}.$$

The chromatic shift at the image position H' is given by Eq. 4.31

$$\delta \approx H'(F\nu(S^{-1} - Z^{-1})^{-1} = -0.032 \text{ mm}.$$

4.7 For a wedge with apex angle α made of glass with refraction index n and Abbe number ν, derive the relation between the refraction angle and the wavelength.

4.8 Using (4.38), derive the expressions (4.40) and (4.41).

References

1. V. Mahajan, *Aberration Theory Made Simple* (SPIE, 1991)
2. W.T. Welford, *Aberrations of Optical Systems* (IOP Publishing, Adam Hilger Series, 1986)

Chapter 5
Introduction to Optical Instruments

Abstract The human eye is the oldest known optical instrument, existing long before the formal science of optics was established. Its structure and basic parameters—such as resolution and color sensitivity—are described, together with the most common vision defects, including myopia (nearsightedness) and hyperopia (farsightedness). The structure of elementary optical instruments composed of two and more components is introduced. This includes systems such as Galilean and Keplerian telescopes, telephoto and wide-field lenses, microscopes, viewfinders, rangefinders, and telescopes. Mathematical expressions defining the fundamental parameters of these systems are derived.

5.1 The Human Eye

Human vision is the primary source of information about the surrounding world. The visual system includes the eyes, the retina, the neural connections to the brain, and the brain itself [1]. In some literature sources the eye is even regarded as part of the brain.

A schematic diagram of the human eye is shown in Fig. 5.1. Protective structures such as eyelids and eyelashes are not depicted, although they play an important role in protecting the eye and also contribute to facial appearance.

The outer shell of the eye is formed by the opaque white sclera, about 1 mm thick. At the center, the sclera becomes transparent and forms the cornea, the eye's outer optical surface. The cornea has an optical power of about 43 diopters, providing nearly two thirds of the total focusing power of the eye. The remaining power is supplied by the crystalline lens, which adjusts its curvature to achieve focusing (accommodation). Since the cornea has a fixed spherical shape, the static optical power of the eye depends strongly on its curvature. Incorrect curvature leads to defocus, resulting in nearsightedness or farsightedness, while more complex distortions cause classical aberrations such as astigmatism and coma. Surgical treatments have been developed to reshape the cornea and improve vision acuity, most commonly for the correction of myopia and astigmatism. The average corneal thickness is about 0.5 mm, though it varies between individuals.

G. Vdovin, *Elementary Technical Optics*, UNITEXT for Physics,
https://doi.org/10.1007/978-3-032-08626-6_5

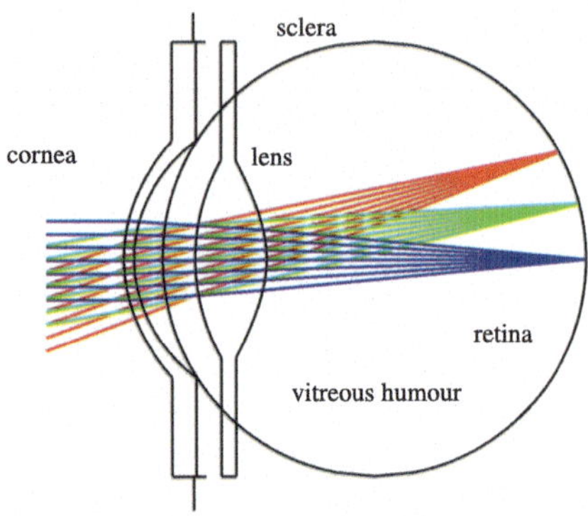

Fig. 5.1 Schematic of the human eye

The crystalline lens serves as the second focusing element. Its aperture is controlled by the iris. The pupil diameter is about 6 mm in dim light and contracts to about 1 mm in bright light. The focusing power of the eye changes through the action of the ciliary muscles: the lens thickens for near focus and relaxes for distant focus. The chamber between the cornea and crystalline lens is filled with aqueous humour, while the space behind the lens is filled with the gel-like vitreous humour (corpus vitreum).

The retina, the light-sensitive part of the eye, lines the back surface. Incoming light must pass through several cell layers before reaching the photoreceptors at its base. The retina, about 250 μm thick, consists of multiple layers: a protective membrane adjacent to the vitreous, a layer of axons carrying neural signals, the ganglion cell layer, layers of bipolar and horizontal cells, and finally the photoreceptor layer of rods and cones. Behind the photoreceptors lies the pigmented layer, containing melanin to absorb stray light and suppress backscattering.

The retina contains about 7 million cones and about 120 million rods. Rods are highly sensitive to light but lack color discrimination. Their spectral sensitivity peaks at 550–565 nm, corresponding to green light reflected by vegetation under solar illumination. Cones are roughly ten times less sensitive but enable color vision. Three types of cones respond to blue, green, and red light. Cones are concentrated near the visual axis, with maximum density in the fovea, a specialized region about 5° off-axis, with an angular size of about 1.2° (0.4 mm in diameter). The fovea contains a dense mosaic of ~25,000 cones, each subtending about 0.5 arcmin, and each directly connected to the optic nerve, enabling the highest angular resolution. Rod density is low at the center but increases toward the periphery. On average, the retina contains about 160,000 photoreceptors per mm^2.

Approximately one million axons connect the retina to the brain. Each foveal cone has a dedicated axon for maximum resolution, while peripheral photoreceptors connect through more complex pathways involving bipolar and horizontal cells. Several rods and/or cones may feed into a single bipolar cell, while one photoreceptor may connect to multiple bipolar cells. Horizontal cells provide lateral connections, forming "receptive fields" that preprocess and compress visual information before transmission to the brain. This structure enhances sensitivity to motion while reducing sensitivity to static images.

The distribution of receptors explains the dependence of vision on illumination. In bright light, cones in the fovea dominate, providing high resolution and color sensitivity. In dim light, rods dominate, yielding monochromatic vision with reduced resolution. This is summarized by the proverb: "All cats are black in the dark."

Accommodation and resolution

In its relaxed state, the healthy human eye is focused at infinity. A general term for failure of the eye to focus properly is *ametropia*. When the eye focuses only at shorter distances and cannot adjust to infinity, the condition is called *myopia* (nearsightedness). When the eye focuses "behind" infinity, the condition is called *hyperopia* (farsightedness).

A healthy eye can accommodate by changing the posterior curvature of the crystalline lens, thus shifting focus from infinity to near distances. This process is controlled by the ciliary muscles. In youth, the accommodation range may reach up to 10 D. With age, the elasticity of the lens decreases, and the accommodation range diminishes, eventually approaching zero. At advanced age, hyperopia often develops in parallel with the loss of accommodation.

Refractive errors due to myopia or hyperopia can be corrected with spectacles, contact lenses, ocular implants, or corneal surgery, including laser treatments. The optical power of the correction must compensate the refractive error of the eye. For example, if the refractive error is +3 D (indicating myopia), then corrective lenses of -3 D must be prescribed.

The human eye also adapts to varying illumination by adjusting the iris aperture between $\sim$6 mm in low light and $\sim$1 mm in bright light. This iris accommodation influences resolution: at small pupil diameters, resolution is diffraction-limited, while at large diameters it is limited by ocular aberrations. For a pupil diameter of 4 mm, the numerical aperture of the eye is $A \approx 0.1$. At $\lambda = 550$ nm, the diffraction-limited linear resolution is 5.5 μm, corresponding to about 0.3 mrad or 1 arcmin in angular resolution.

Model eye

In many optical design tasks it is useful to employ a simplified model of the human eye. Such models capture only the essential parameters, since detailed anatomy varies among individuals. The designers of the Zemax software [2] recommend a "reduced" eye for paraxial calculations, consisting of a single refracting surface with 60 D of power and a refractive index of 4/3. This yields a surface radius of 5.55 mm and an axial length of 22.22 mm. The model ignores retinal curvature and is therefore valid only in the paraxial region. Parameters of a more detailed Zemax model eye are listed in Table 5.1.

Two earlier models, due to Gullstrand [3] and Verbitsky [4], are summarized in Table 5.2.

Table 5.1 Parameters of a detailed model of the human eye

Surface	ROC, mm	Thickness, mm	Glass code
Cornea	7.77	0.52	136667
Aqueous	6.7	1.5	132847
Aqueous	11	1.6	132847
Iris (stop 1 to 7 mm)	∞	0.1	132847
Lens	10	3.7	140939
Vitreous	–6	16.58	132673
Retina	–11		

Table 5.2 Parameters of model eyes

Parameter	Gullstrand	Verbitsky
Length of the eye, mm	24	23.4
ROC of the cornea	7.7	6.8
Refractive index	1.34	1.4
ROC of the retina	10.5	10.2
Back focal length	22.785	23.8

More detailed optical parameters of the human eye, according to [4, 5], are presented in Table 5.3.

The accommodation of a model eye can be simulated by adjusting the radii of curvature of the crystalline lens surfaces in Table 5.1.

5.2 Tele and Wide-Field Two-Component Systems

If a single lens is used to create an imaging system, the total length of the optical system must be of the order of the focal length. This is not always acceptable, especially when the focal length is very short or very long. Two-component optical systems make it possible to design both long-focus and short-focus systems with practically acceptable dimensions.

To establish the relation between the focal length and the configuration of a two-component system, consider two thin lenses with focal lengths F_1 and F_2, separated by a distance L. Similar to the case of a single lens described in Sect. 3.3, we can calculate the positions of the principal planes of the two-lens system, T_1 and T_2, the focal length F, and the distance from the last component to the back focal plane S_b.

The layout of a two-component lens begins with the requirements for the distance between the last component and the focal plane (back focus) S_b, the distance between

Table 5.3 Optical parameters of the human eye

Parameter	Infinity accomm.	Max accomm.
Refractive index		
Cornea	1.376	
Aqueous and vitreous humour	1.336	
Crystalline lens	1.386	
Crystalline lens kernel	1.406	
Distance to cornea front		
Cornea back surface	0.5	
Front cryst. lens	3.6	3.2
Back cryst. lens	7.2	
ROC, mm		
Front cornea	7.7	
Back cornea	6.8	
Front cryst. lens	10	5.33
Back cryst. lens	–6	–5.33
Retina	–10.5	
Focusing power, D		
Front cornea	48.83	
Back cornea	–5.88	
Front cryst. lens	5	9.375
Back cryst. lens	8.33	9.375
Lens kernel	5.985	14.94
Cornea focusing power	43.05	
Cryst. lens focusing power	19.11	33.06
Total focusing power	58.64	70.57
Back focus, mm	22.785	18.93

the components L, and the focal length F. The focal lengths of the first and second components of such a system are then given by:

$$F_1 = \frac{LF}{F - S_b},$$
$$F_2 = \frac{LS_b}{S_b - F + L}. \tag{5.1}$$

The positions T_1 and T_2 of the first and second principal planes with respect to the lens of focal length F_1 are:

$$T_1 = \frac{LF_1}{F_1 + F_2 - L},$$
$$T_2 = \frac{L(F_1 - L)}{F_1 + F_2 - L}. \quad (5.2)$$

Expressions (5.1) are useful for the layout of telephoto and retrofocus systems:

- Telephoto lenses, composed of two separate components as illustrated in Fig. 5.2, are used for imaging remote objects. The action of such a system is illustrated with an object at infinity. To keep the system length short compared to its focal length, it is advantageous to make both the distance from the last component to the focal plane S_b and the system length L much smaller than F. This means $F > L$ and $F > S_b$.
  Consider an example in which the total system length, from the front component to the focal plane, is five times shorter than the focal length. Assume $S_b = 1$, $L = 1$, and $F = 10$. Using expressions (5.1) we obtain $F_1 = 10/9$ and $F_2 = -1/8$. From (5.2) we find $T_1 = -80$ and $T_2 = -8$. Although the system is very short, the back principal plane is positioned far in front of the first component. This arrangement provides a long focal length while keeping the overall system compact.
- Retrofocus lenses, composed of two components as illustrated in Fig. 5.3, are used when the distance from the last component to the focal plane is comparable to, or much longer then the focal length of the system. A typical example of such a lens is a wide-field lens for a single-lens reflex (SLR) camera. In an SLR the space between the last surface of the lens and the focal plane is occupied by a tilted mirror. This mirror is required to form the viewfinder image for focusing. The mirror size is comparable to the image frame, forcing the back focal distance to be large. Retrofocus designs are therefore used for wide-field lenses when the condition $F < S_b$ must be satisfied.

Figure 5.4 illustrates that lenses with focal lengths of 300 and 17 mm can be realized within nearly identical package dimensions.

Consider a retrofocus lens with $F = 1$, $L = 1$, and $S_b = 5$. Using expressions (5.1) we obtain $F_1 = -1/4$ and $F_2 = 1$. From 5.2 we find $T_1 = 1$ and $T_2 = 5$.

The focal length of an imaging system formed by a positive lens can also be extended by placing a negative lens close to the system focus. A negative Barlow

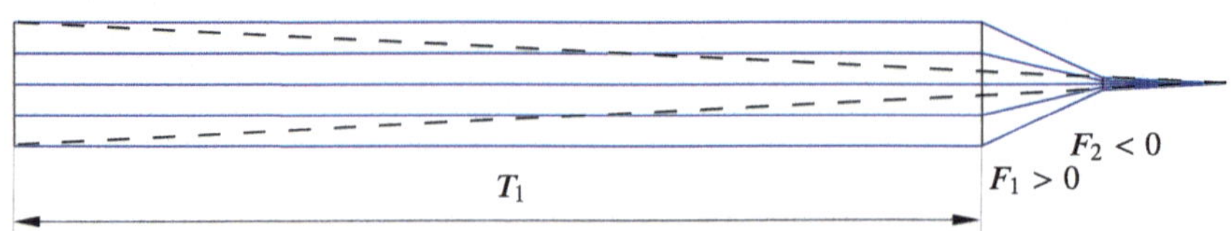

Fig. 5.2 Ray tracing in a telephoto lens. The back principal plane is positioned at the distance T_1 in front of the front element. The telephoto lens can be represented as a combination of a positive imaging lens with short focal length $F_1 > 0$, and a negative Barlow lens $F_2 < 0$ inserted in front of the system focus

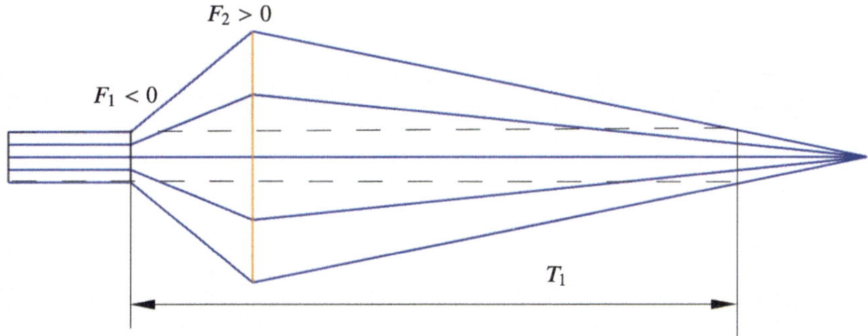

Fig. 5.3 Ray tracing in a retrofocus lens. The back principal plane is positioned at the distance T_1 behind the front element, close to the focal plane

Fig. 5.4 The lens on the left combines refraction and reflection (a catadioptric design) to achieve a focal length of 300 mm in a compact package. The lens on the right with a focal length of 17 mm employs retrofocus design. Both lenses are intended for use with the same system camera

lens with focal length f, positioned at distance L in front of the original focal point, produces a new focal length F_B:

$$F_B = \frac{F \cdot f}{L - f}. \tag{5.3}$$

Example: A Barlow lens with $f = -100$ mm is placed at a distance $L = 50$ mm in front of the focus of a telescope with $F = 1000$ mm and $F\# = 6$. Calculate the resulting focal length F_B and the new $F\#_B$.

Solution: Using Eq. 5.3 we obtain

$$F_B = \frac{-100 \cdot 1000}{50 - 100} = 2000. \tag{5.4}$$

The new focal ratio is $F\#_B = F\# \cdot \frac{F_B}{F} = 12$.

In practice, the solutions obtained with (5.1) are often difficult to implement with only two components. This is due to limitations related to aberrations, field of view, and component size. As a result, two-component designs serve primarily as conceptual layouts for more complex systems.

5.3 Afocal Systems

If two components have focusing powers $D_1 = F_1^{-1}$ and $D_2 = F_2^{-1}$, and separation L, the expression for the focusing power of the system can be written as:

$$D = D_1 + D_2 - L D_1 D_2. \tag{5.5}$$

The special case of

$$F_1 + F_2 = L \tag{5.6}$$

represents the so-called *afocal system* (Fig. 5.5).

For an object at infinity with angular size φ, the afocal system forms its image at infinity with angular size $\psi = M_a \varphi$, where

$$M_a = -\frac{F_1}{F_2} \tag{5.7}$$

is the angular magnification.

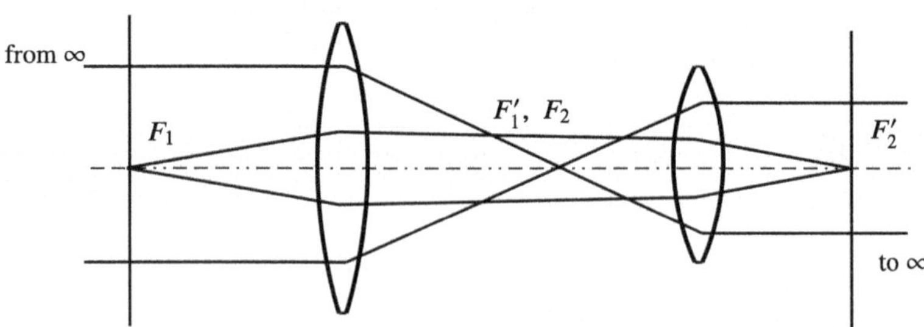

Fig. 5.5 Afocal system formed by two components with focal lengths F_1 and F_2. Infinity is imaged to infinity, and the entrance pupil in plane F_1 is imaged to the exit pupil in the plane F_2'

If both lenses of the afocal system are positive, we can position the object in the focal plane of the first lens. The system then forms the real image of the object in the focal plane of the second lens. These focal planes are optically conjugated, because the two-lens system preserves not only the geometrical correspondence between the object and the image, but also the ray directions in the image plane are matched with the rays in the object plane. This property makes it indispensable for re-imaging and replicating pupil positions. It should be noted that only the Keplerian configuration can be used for pupil re-imaging, since the Galilean configuration does not form a real image of the exit pupil.

The linear (transversal) magnification between the two conjugated planes in the afocal system is given by

$$M_t = -\frac{F_2}{F_1}. \tag{5.8}$$

Afocal systems are frequently used as beam expanders in laser technology. If the afocal system has magnification M, the input beam is collimated and has the diameter A_1, while the exit collimated beam has the diameter A_2. The ratio of beam diameters is defined by the system magnification:

$$\frac{A_2}{A_1} = \left|\frac{F_2}{F_1}\right| = |M_t|. \tag{5.9}$$

Then the divergence of the exit beam is reduced by a factor of M_t. The angular magnification of the afocal system can be practically estimated as the ratio of the diameter of the input and exit pupils, as shown in Fig. 5.6.

Angular magnification is an essential parameter in the design of telescopes. There are two basic configurations of a telescope that satisfy the afocal condition (5.6):

- *Keplerian system*, formed by two positive lenses with $F_1 > 0$ and $F_2 > 0$. The human eye sees an inverted image through such a system. Since the first component forms a real image conjugated to infinity, transparent reticles (printed on thin flat glass plates) can be superimposed with the object image. Such reticles are used,

Fig. 5.6 A Keplerian monocular with a magnification of 7 and an entrance pupil diameter of 18 mm is shown. The central and right images demonstrate that the magnification can be estimated as the ratio of the entrance to the exit pupil diameters

for example, in measurement microscopes and rifle sights. The introduction of an additional positive lens inside the Keplerian system, re-imaging the focal plane of the objective lens, produces an upright image. Also prism systems are frequently used in Keplerianian systems to obtain the upright image.

- *Galilean system*, formed by a pair of positive and negative lenses with $F_1 > 0$ and $F_2 < 0$. The magnification M_t is positive, so the system forms an upright image. However, there is no object-conjugated plane in the system. A Galilean system therefore lacks the possibility to use focal plane reticles superimposed onto the object image.

Both systems form a real image at infinity. The human eye, accommodated to infinity, can observe this image if the pupil of the eye is positioned at the exit pupil of the afocal system. For the Keplerian system, the exit pupil is located outside the system. For the Galilean system, the exit pupil is formed inside the system, and therefore cannot be directly coupled to the human eye. For this reason, Galilean systems are less suited for instrumental applications involving a human observer.

The Keplerian system has two drawbacks. The image is inverted, and the rays in the intermediate image plane are not telecentric. The inverted image can be corrected by introducing an inverting $4F$ system, which flips the real image with magnification $M_i = -1$. However, the introduction of the inverting element does not solve the problem of ray aiming. This problem is solved with a *field lens* placed exactly in the image plane. The field lens positioned in the image plane does not change the image itself. Instead, it redirects the ray pencils and makes them telecentric, as shown in Fig. 5.7.

The following requirements must be considered in the design of a Keplerian system for observation or aiming purposes:

- The diameter of the exit pupil should match the diameter of the human eye pupil, usually in the range of 2.5–5 mm.
- The position of the exit pupil should allow comfortable placement of the observer's eye at a safe distance from the last surface of the ocular lens. The position of the exit pupil depends on the parameters of the inverting lens and the field lens.

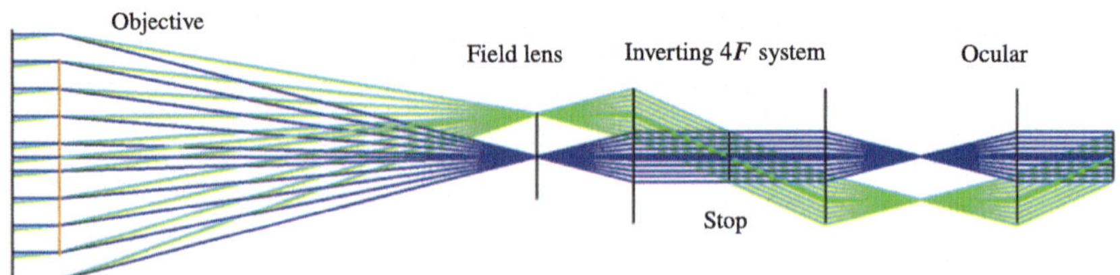

Fig. 5.7 Keplerian system with field lens and inverting $4F$ system. If the focal length of the field lens equals that of the objective, the stop position is conjugated to the aperture of the objective. The stop is also conjugated to the eye pupil. Measurement reticles can be placed either in the plane of the field lens or in the secondary focal plane formed by the inverting $4F$ system

- The magnification, defined by the ratio of the entrance pupil (usually the first lens of the system) to the exit pupil, should be matched to the observer's ability to hold the system stable to prevent image shake. For handheld binoculars, the comfortable angular magnification is limited to about 3–10. For tripod-mounted or actively stabilized systems, the magnification can be higher.
- Since the longitudinal magnification in the conjugated planes equals the square of the linear magnification, binoculars and observation tubes produce the effect of visually "compressed" space, which distorts the normal visual perspective.

Keplerian afocal systems are widely used in technical and laser optics for pupil re-imaging.

5.4 Loupe and Microscope

The standard accommodation distance of the human eye is $l_a = 25$ cm. The focal length of the eye F_e can be taken as 20 mm. Then, the lateral magnification is given by $M_t \approx F_e/l_a \approx 0.08$.

A positive lens with focal length F_l, placed in front of the eye as shown in Fig. 5.8, reduces the observation distance to F_l, while keeping the eye accommodated at infinity. Then, the object appears l_a/F_l times larger than if it were observed with the naked eye from the standard distance of 25 cm. The ratio

$$M_l = 1 + \frac{l_a}{F_l} \tag{5.10}$$

is called the visual magnification of the loupe.

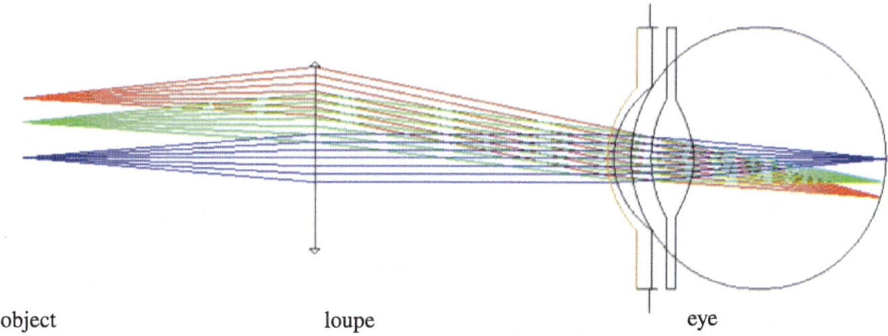

Fig. 5.8 Human eye with a magnifying glass (loupe)

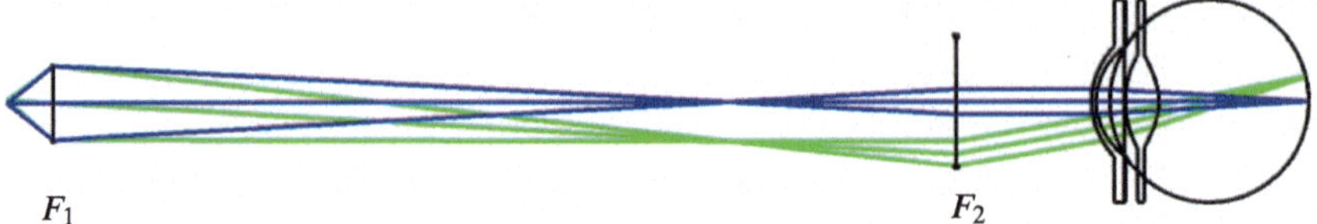

Fig. 5.9 Human eye viewing through a microscope

A single loupe with high magnification is very small, because a short focal length implies a small lens size. The smallest loupe, used in Antoni van Leeuwenhoek's microscope, had a diameter of about one millimeter. It was reported to be difficult to use, as it had to be placed very close to the eye and also had a very small pupil.

An optical microscope, shown in Fig. 5.9, achieves very high magnification by using an additional objective lens that forms a real intermediate image of the object. This intermediate image is then observed through the ocular, in the same way as with a loupe. The distance L_t at which the objective forms the intermediate image is called the tube length. This parameter, typically in the range 160–200 mm, differs between microscope manufacturers. The standard Zeiss tube length is 160 mm, Leica and Nikon use 200 mm, and Olympus uses 180 mm. To avoid confusion, microscope objectives are marked with the tube length on their barrels. According to Eq. 5.9, the magnification of the objective is

$$M_1 = -\frac{L_t - F_1}{F_1}\bigg|_{F_1 \ll L_t} \approx -\frac{L_t}{F_1}. \tag{5.11}$$

The magnification of the ocular is given by Eq. 5.10: $M_2 = \frac{250\ \text{mm}}{F_2}$.

The total magnification of the microscope, when coupled to the human eye, is the product of the ocular and objective magnifications: $M = M_1 M_2$. For different applications, microscope objectives can be designed with magnifications ranging from 1 to 100.

Microscope objectives are corrected to minimize aberrations when used at their specified geometrical magnification, as marked on the objective barrel. This usually corresponds to the conjugation between the sample and the image plane at the tube lens distance. In this configuration, any image sensor placed in the image plane instead of the ocular will record the corrected and magnified image.

Some objectives are corrected for building the image at infinity. To form an image on a sensor at a finite distance, these objectives require an additional lens, called a tube lens, which forms a $2F_1 + 2F_2$ system to create the real image. The geometrical magnification of a system with a tube lens of focal length F_2 and an objective of focal length F_1 is given by Eq. 5.8. When F_2 is equal to the nominal tube length, the system magnification equals the magnification of the objective. Microscope objectives are marked with magnification, numerical aperture, and cover glass thickness. The cover glass placed over the sample introduces spherical aberration according to Eq. 4.10, so objectives are designed and manufactured pre-corrected for this effect.

Infinity-corrected objectives provide greater freedom in designing complex microscopy systems. By using beamsplitters and different tube lenses, a single infinity-corrected objective can create images with different magnifications in separate image planes. Infinity-corrected objectives can also be adapted for epi-illuminated imaging, where illumination and observation are performed through the same objective. Epi-fluorescence microscopy uses excitation light at a specific wavelength delivered through the imaging objective, while observation is performed at a longer emission wavelength through the same objective.

A cellphone can be adapted for recording images from a standard microscope, by placing it at the exit pupil in place of the human eye. Then the exit pupil of the microscope should match the entrance pupil of the cellphone camera, which is not always the case. Another option is to use the phone's lens as a tube lens. In this case, the micro-objective must have a focal length shorter than that of the phone lens to achieve magnification larger than $|M_t| > 1$. Small glass spheres with very short focal lengths are often used as objective lenses in simple cellphone-based microscopes. Although such systems suffer from strong aberrations, they are extremely simple to implement and may be useful for low-cost microscopy applications.

5.5 Köhler condenser

The illumination system of a microscope must satisfy strict requirements:

- provide uniform illumination within a restricted region of diameter D in the specimen plane,
- deliver illumination with numerical aperture A matched to that of the microscope objective.

The simplest illumination scheme employs an extended uniform source of diameter d imaged directly onto the specimen plane. However, this approach superimposes the source structure onto the specimen, creating artifacts. To avoid this, the source image is often projected in a defocused manner. Nevertheless, such a method cannot provide large numerical apertures, truly uniform light sources are scarce, and independent control of the illumination diameter D and the numerical aperture A is limited.

Köhler illumination satisfies these requirements, making the Köhler condenser the standard in modern microscopy. Its schematic is shown in Fig. 5.10. The light source of size d is placed at the focal point of lens L_1, producing an illumination beam with numerical aperture $\sim d/(2F_1)$ in the focal plane of L_1, where the field stop a_1 is located.

The focal plane of L_1 is reimaged by the afocal system of L_2 and L_3 onto the specimen plane. The aperture stop a_2, positioned in the common focus of L_2 and L_3, controls the numerical aperture of illumination. With all stops fully open, the

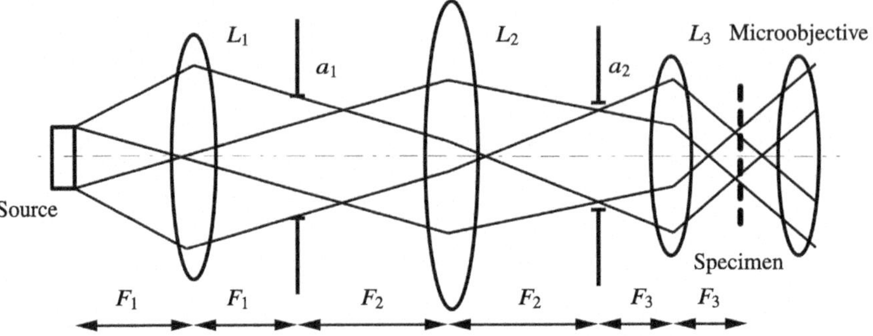

Fig. 5.10 Köhler illumination

numerical aperture in the specimen plane is defined by the physical size of the source:

$$A \approx \frac{d}{2F_1} \frac{F_2}{F_3}.$$

In practice, the numerical aperture is adjusted by a_2

$$A \approx \frac{a_2}{2F_3},$$

while the illuminated field is set by a_1

$$D = \frac{a_1 F_3}{F_2}.$$

To illustrate the basic principles, these formulas are given in paraxial approximation for small numerical apertures $A \ll 1$. Trigonometric relations and exact ray-tracing is recommended for condenser layout with numerical apertures $A \gg 0$. Thus, the Köhler condenser provides precise and independent control of the illumination field and numerical aperture in microscopy and can be implemented for large numerical apertures, to match these of modern micro-objectives.

5.6 Albada Viewfinder

The Keplerian system has an intermediate image plane in the focus of the first component, which can be used for placing reticles superimposed over the field of view. The Galilean system does not have any intermediate plane conjugated to infinity, therefore it seems impossible to superimpose any raster image over the observed object.

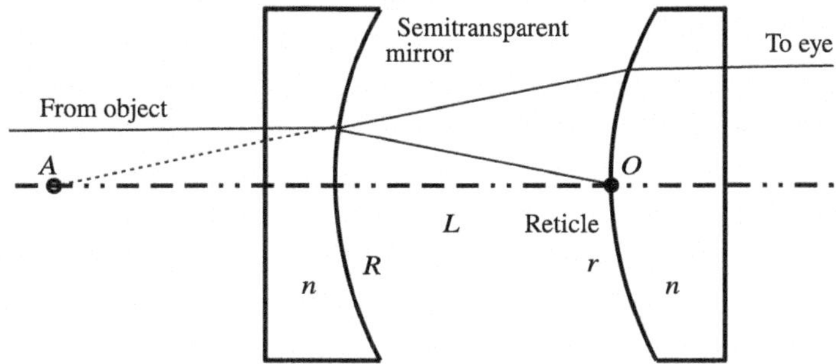

Fig. 5.11 Schematic of the Albada viewfinder

An original solution to this problem was proposed by Lieuwe E. V. Van Albada; see, for instance, US patent US1678493A.

The schematic of the Albada viewfinder is shown in Fig. 5.11. The system includes two separate arms, formed by the surfaces of the same components:

- In the object arm, the front and back lenses, separated by an air gap L, form an afocal Galilean system that traces the ray from the object at infinity to the eye.
- The reticle arm is formed by a reticle printed on the convex surface of the second lens with radius r, and a mirror created by a semi-reflective coating on the concave surface of the first component with radius of curvature R. The reticle, reflected in the mirror, is observed through the second lens as projected into infinity. Since both arms are visible to the eye, the image of the reticle in point O is superimposed with the image of the object.

To find the parameters of the viewfinder, we need to determine R and r as functions of L and n, under the condition that both the object at infinity and the reticle are projected to infinity in the image space.

In the object arm, the focus of the first lens must coincide with the focus of the second lens in point A. Using the simplified form of the lensmaker's equation (3.18):

$$\frac{R}{n-1} + L = \frac{r}{n-1}. \tag{5.12}$$

In the reticle arm, the reticle at point O is imaged by the concave mirror with radius R into point A. The focal length of a concave mirror equals $R/2$, so

$$\frac{1}{2/R - 1/L} + L = \frac{r}{n-1}. \tag{5.13}$$

The nontrivial solution of the system (5.12) and (5.13) is:

$$R = L(3-n), \quad r = 2L. \tag{5.14}$$

The angular magnification of the object arm is defined as

$$M = \frac{R}{r} = \frac{3-n}{2}. \tag{5.15}$$

To find the relation between the size of the reticle and its observed angular size in the reticle arm, we calculate the focal length of the combination of the concave mirror, with focal length $F_1 = R/2$, and the lens, with focal length $F_2 = r/(n-1)$. Using (5.14) and (5.5), we obtain:

$$F = \frac{F_1 F_2}{F_1 + F_2 - L} = \frac{2L}{n+1}. \tag{5.16}$$

The observed angular size of the reticle is

$$\varphi = \arctan\left[\frac{H}{F}\right] = \arctan\left[\frac{H(n+1)}{2L}\right], \tag{5.17}$$

where H is the linear size of the reticle.

Expressions (5.14), (5.15), (5.16), and (5.17) define the parameters of the Albada viewfinder formed by plano-concave and plano-convex lenses. Additional degrees of freedom can be introduced by bending the flat surfaces of the objective and ocular lenses.

5.7 Triangulation Rangefinder

Figure 5.12 illustrates the principle of coincidence and binocular rangefinders. Both configurations are based on the triangulation principle. The object is observed in two different arms separated by the base B. The difference between the observation directions is defined by the angle α. Then, the distance to the object L, for $\alpha \ll 1$, is given by

$$L = \frac{B}{\alpha}. \tag{5.18}$$

In the coincidence rangefinder the angle α is measured by observing two superimposed images of the object. By rotating the alignment mirror in one of the arms, two images can be aligned. We assume both arms have the same magnification.

The principle of the binocular rangefinder is very similar. Each arm has a mark projected on the object. The object is observed with both eyes, and the distance is measured by changing the angle in one arm to make the images of two marks coincident.

The angular precision $\Delta\alpha$ of subjective coincidence of two images depends on the angular resolution of the eye. Usually the alignment error is a fraction of the

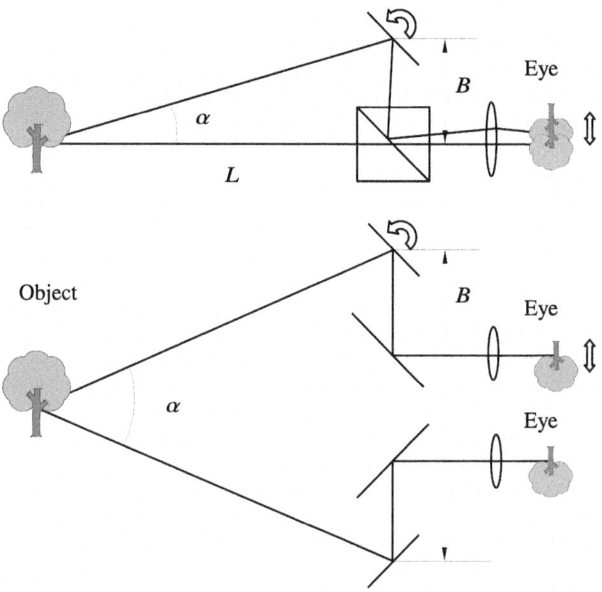

Fig. 5.12 Coincidence (top) and binocular (bottom) rangefinders

angular resolution, but it depends on the observer and the conditions. Afocal systems with magnification $M > 1$ are used in rangefinders to improve the precision. Afocal systems with $|M| < 1$ are used to extend the field of view. Using 5.18, the range error ΔL can be expressed as

$$\Delta L \approx \frac{dL}{d\alpha}\frac{\Delta\alpha}{M} = -\frac{B}{\alpha^2}\frac{\Delta\alpha}{M} = -\frac{L^2}{B}\frac{\Delta\alpha}{M}. \tag{5.19}$$

It follows from (5.19) that high precision of the rangefinder is achieved with large base B and magnification M.

The triangulation principle has been widely used in point-and-shoot cameras with automatic rangefinders. Active rangefinders project a light spot on the object and calculate the distance from two images of the spot. Using narrowband modulated illumination, these systems achieve high reliability even under bright and noisy ambient light.

Another approach to distance measurement consists in emitting very short light pulses with duration τ and measuring the return flight time t for the pulse reflected by the object. The distance L and the measurement error ΔL are given by

$$L = 0.5tc, \quad \Delta L = 0.5\tau c, \tag{5.20}$$

where c is the speed of light.

5.8 Time of Flight and Phase Rangefinders

The distance can also be measured by modulating a continuous light beam with frequency f and period $T = 1/f$. The phase shift ϕ between the outgoing and returned beams is proportional to the distance. The delay of the reflected wave is $t = 2L/c$, then the phase shift is $\phi = 2\pi t/T$. Combining these expressions, we obtain

$$L = \frac{c\phi}{4\pi f} + k\frac{c}{f}, \tag{5.21}$$

where $k = 0, 1, 2\ldots$ is the phase order. To eliminate the ambiguity we assume $k = 0,\ \ \phi = 2\pi$. Then the modulation frequency that matches the distance range $0\ldots L$ to the phase delay range $0\ldots 2\pi$ is

$$f = \frac{c}{2L}. \tag{5.22}$$

For $L = 1$ km the modulation frequency should be 150 kHz. For a 10 m range the modulation frequency should be 15 MHz.

The precision of a phase rangefinder can be further increased by using phase orders $k \geq 1$. To determine k, the phase is measured at two closely chosen modulation frequencies $f \gg c/(2L)$ and $f_1 = f \pm \Delta$. The phase order in the range $k < f/\Delta$ can then be determined from two phase measurements at these frequencies.

Fine displacements are measured with interferometric setups, such as the Michelson interferometer described in Sect. 6.1. Interferometric measurements provide positioning precision in the nanometer range. They are used in lithography for precise alignment of wafer position and orientation in lithographic steppers and scanners.

5.9 Two-Mirror Telescopes

The angular resolution and light-gathering power of a telescope are both determined by its aperture. Refracting telescopes that rely on lenses are limited technologically, since the fabrication of large, aberration-free lenses is extremely difficult. For this reason, reflecting telescopes are the preferred solution for large apertures.

A simple reflector using a single parabolic mirror provides good on-axis resolution but suffers from a long size and off-axis aberrations. Two-mirror systems, consisting of a primary and a secondary mirror, offer greater design flexibility.

The principal two-mirror telescope layouts are shown in Fig. 5.13. Cassegrain implements a mirror-based variant of the Galilean scheme, while Gregorian can be designed in afocal configuration with two parabolic mirrors to implement the Keplerian scheme. In all basic two-mirror configurations, a parabolic primary mirror of focal length F_1' is combined with a secondary mirror positioned at distance S_1' from the primary focus, with total distance d between mirrors. Then the equivalent

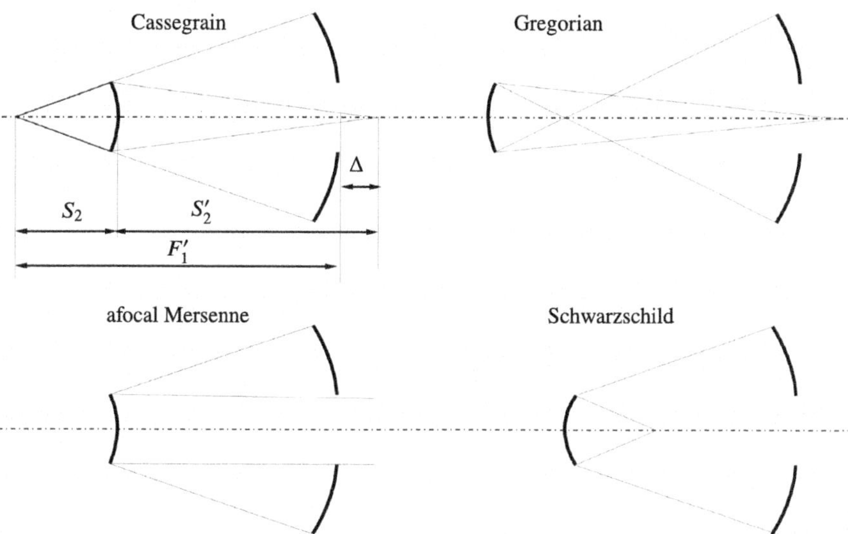

Fig. 5.13 Basic two-mirror telescope configurations

focal length of the two-mirror system is F_{eq}. The secondary mirror re-images the primary focus to the position S_2', measured from the secondary mirror. With known F_{eq}, d, and S_2', the radii of the primary and secondary R_1 and R_2 can be calculated using formulas (5.1). The conic constant of the secondary is defined by (4.14), while the conic of the primary is $e_1^2 = -1$.

The Mersenne telescope represents a special case of infinity-conjugated afocal system. It produces collimated output that, with the help of beam splitters, can be used to feed multiple secondary instruments such as imagers and spectrometers.

These fundamental layouts eliminate spherical aberration and provide aberration-free on-axis imaging, though they remain affected by coma and astigmatism. Modern two-mirror variants include the Ritchey–Chrétien telescope, employing hyperboloidal primary and secondary mirrors, which corrects both coma and spherical aberration and is widely used in professional astronomy, including the Hubble Space Telescope. Other designs such as the Dall–Kirkham with a spherical secondary, and shortened Schwarzschild layouts, offer alternative solutions for off-axis correction.

5.10 Conclusion

The human eye functions as a compound optical instrument combining a fixed-power cornea with a variable-power crystalline lens. Its accommodation range is governed by the elasticity of the lens and controlled by the ciliary muscles. The resolution is limited by diffraction, ocular aberrations, and the photoreceptor density. Myopia and hyperopia represent typical focus error of the eye, usually corrected by spectacles. A number of eye models is given for simulation of the visual performance.

By controlling focal lengths of the components and their separation in a two-component optical system, telephoto and retrofocus configurations can be designed. Telephoto lenses provide long effective focal length in a short package, while retrofocus systems enable wide fields with long back focus. The Barlow lens is used to modify the effective focal length.

Afocal systems are formed by two components with common focus. The angular magnification of an afocal system is given by the ratio of the focal lengths of the components for both Keplerian and Galilean configurations. Keplerian instruments allow reticles at conjugated planes and external position of the exit pupil. Galilean instruments produce compact upright images but lack intermediate image planes.

The loupe increases visual magnification by reducing the effective observation distance. The microscope combines an objective and ocular, extending the magnification range, compared to loupe. Infinity-corrected objectives with tube lenses enable modular microscope architectures and epi-illumination methods.

In microscopy, direct imaging of an extended illumination source onto the specimen leads to artifacts and poor control of illumination parameters. Köhler illumination overcomes these drawbacks by re-imaging the source through relay optics, with field and aperture stops providing independent adjustment of illuminated area and numerical aperture.

In astronomy, the single parabolic mirror offers high on-axis resolution but suffers from aberrations and results in impractically long systems. Two-mirror schemes, such as Gregorian, Cassegrain, Ritchey–Chrétien, and Dall–Kirkham, extend the design space by using aspherical secondary mirrors to shorten the system, and suppress aberrations.

5.11 Problems

5.1 Grandma Gigi has accommodation range from infinity to 2 m. For reading she wears glasses with focusing power of 2 diopter. Estimate the front and back boundaries of her accommodation range when she wears glasses.

Solution (5.1). The far boundary is defined by the focusing power of glasses D [diopters] and equals to

$$L_{far} = \frac{1}{D} = 0.5[\mathrm{m}].$$

The focusing power of eye D_e, focused to $L = 2$ m equals $D_e = 1/L = 0.5$. The near boundary equals to

$$L_{near} = \frac{1}{D + D_e} = 0.4[\mathrm{m}].$$

5.2 Ammeotropic eye can not focus to infinity, the farthest distance of sharp vision is $L = 50$ cm. Is the eye myopic or hyperopic? Estimate the refraction error. Suggest prescription of spectacles.

Solution (5.2). The myopic refractive error is $D = 1/L = 2$ d. Spectacles of –2d should be prescribed. Although this method is technically sound, the prescription of spectacles is medical matter, the doctor's advise is always recommended for spectacle prescription!

5.3 In its essence, the Antoni van Leeuwenhoek's microscope is a glass ball loupe. Calculate the visual magnification of the microscope for a ball with a diameter of 2 mm, made of glass with $n = 1.5$, coupled to the eye. Assume the diameter of the working aperture of the ball is stopped to 1 mm, to block the aberrated edge rays.

Solution (5.3). Focal length of the glass ball with $r = 1$ mm and refractive index $n = 1.5$:

$$f = \frac{n \cdot r}{2(n-1)} = \frac{1.5 \cdot 1\,\text{mm}}{2(1.5-1)} = \frac{1.5\,\text{mm}}{1} = 1.5\,\text{mm} = 0.15\,\text{cm}.$$

Magnification:

$$M = \frac{25\,\text{cm}}{f} = \frac{25\,\text{cm}}{0.15\,\text{cm}} \approx 166.67 \approx 170.$$

5.4 Autofocus point and shoot camera has two positions of the lens: focused to the infinity and focused to $a = 1.5$ m. When the camera is re-focused from infinity to 1.5 m, the whole lens is shifted by $b = 1$ mm along its optical axis. Find the focal length F of the lens.

Solution (5.4). According to 2.15, $b(a-b) = F^2$,
then $F = \sqrt{ab - b^2} = \sqrt{0.001 \cdot 1.5 - 1 \cdot 10^{-6}} = 0.0387$ m.

5.5 Design Keplerian telescope with magnification $M = -2$, length 10 cm, made of a single piece of BK7 glass.

Solution (5.5). The refraction index of air $n = 1$, the refraction index of BK7 $n_1 \approx 1.519$. The collimatd beam is focused inside glass rod at the distance S_1 by the first surface, and then collimated again by the second surface positioned at the distance S_2 from the focus. To achieve $M = -2$, S_1 should be be twice larger than S_2, and their sum should be equal to the total length of 0.1 m:

$$\frac{S_1}{S_2} = 2,$$
$$S_1 + S_2 = 0.1,$$

with solution $S_1 = 0.0666\ldots$, and $S_2 = 0.0333\ldots$ m. Using the Abbe invariant for the first surface, we have:

$$n\left(\frac{1}{R_1} - \frac{1}{\infty}\right) = n_1\left(\frac{1}{R_1} - \frac{1}{S_1}\right), \text{ or } R_1 = \frac{n1 - 1}{n_1} S_1 \approx 0.02277 \text{ [m]}.$$

Then, $R_2 = R_1/M = -0.011388$ m.

5.6 Laser beam is scanned with a scanner shown in Fig. 5.14. The position of the focal point is defined by rotation of wedges around the z axis by angles φ_1, and φ_2 respectively. Find the expression for φ_1 and φ_2 to move the focal spot to the position $(\Delta x, \Delta y)$, if the configuration shown in the figure corresponds to zero deflection, $\varphi_1 = \varphi_2 = 0$. Both wedges have apex angle α and refraction index n (Fig. 5.14).

Solution Maximum deflection Δ_{max} is achieved when $\varphi_2 = \pi$, relative to the position shown in the figure. Then both wedges contribute in the same direction, $\Delta_{max} = 2\alpha(n-1)F$. It is convenient to use polar coordinates, to solve the problem. First we need to turn the wedges in opposite directions by an angle

$$\varphi_1 = \arcsin\left[\frac{\sqrt{\Delta_x^2 + \Delta_y^2}}{2\Delta_{max}}\right]$$
$$\varphi_2 = -\varphi_1, \tag{5.23}$$

to move the focal spot to the point with coordinate $(x, y) = (\sqrt{\Delta_x^2 + \Delta_y^2}, 0)$. Then we need to turn both wedges by an angle $\phi_{1,2} = \arctan(\Delta y/\Delta x)$, to move the focal spot to the point $(\Delta x, \Delta y)$. Finally:

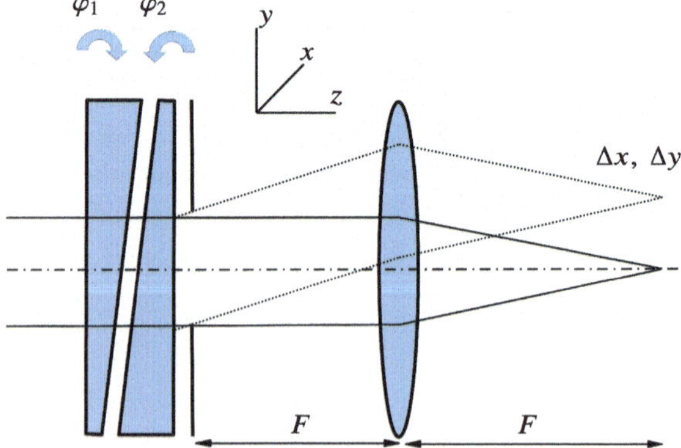

Fig. 5.14 To Problem 5.6

$$\varphi_1 = \arctan\left[\frac{\Delta y}{\Delta x}\right] + \arcsin\left[\frac{\sqrt{\Delta_x^2 + \Delta_y^2}}{2\Delta_{max}}\right]$$

$$\varphi_2 = \arctan\left[\frac{\Delta y}{\Delta x}\right] - \arcsin\left[\frac{\sqrt{\Delta_x^2 + \Delta_y^2}}{2\Delta_{max}}\right] \quad (5.24)$$

5.7 All components of the system shown in Fig. 5.15 are fabricated from BK7 glass with MIL code 517642. The prism has no reflective coating. For an object at infinity $S = \infty$, find the maximum numerical aperture of the lens.

Solution (5.7). The numerical aperture is limited by the TIR condition on the prism hypotenuse:

$$sin(\pi/4 - \arcsin(A/n) > 1/n,$$

with a solution

$$A < nsin(\pi/4 - \arcsin(1/n)).$$

Substituting $n = 1.517$ we obtain $A < 0.1$.

5.8 Figure 5.16 indicates the distances between components in a non-inverting afocal Kepler system formed by 5 positive lenses. Describe the function of each component. Define the focal length of each component. Calculate the angular magnification of the system.

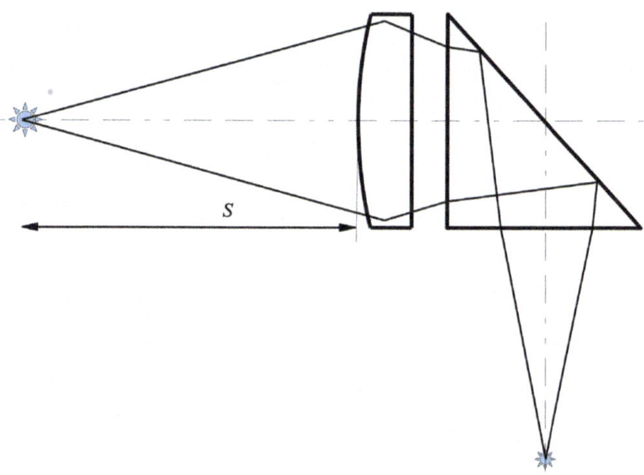

Fig. 5.15 To Problem 5.7

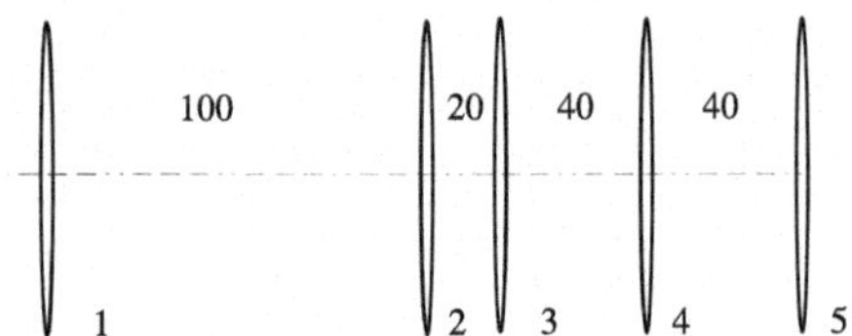

Fig. 5.16 Non-inverting afocal Kepler system, to Problem 5.8

Solution (5.8). First we assume that the lens 1 is the objective, and the focal plane of the lens 1 is coincident with the collective lens 2. Then, $F_1 = 100$. The function of the lens 2 is to aim the off-axis ray cones parallel to the optical axis, which is achieved with $F_2 = 100$. Then, it is logical to assume that the lenses 3 and 4 form a 4F inverting system. Focal length F_3 should be equal to the distance between the second and the third lens: $F_3 = 20$, and, since lenses 3 and 4 should have common focal plane, and the distance between them is 40, we conclude that $F_4 = F_3$. Finally, the lens 5 is the ocular with $F_5 = 20$. Total magnification is positive, thanks to the inverting system: $M_{total} = 100/20 = 5$.

5.9 Thermal deformation has caused the focus of the lens with focal length F to shift to the point $F + \Delta$ along the optical axis. Find the corresponding expression for the wavefront aberration W.

Solution (5.9). The aberration is given by $A \approx \frac{dW}{d\Delta}\Delta$. For $W = \rho^2/(2(F + \Delta))$ we obtain the aberration $A \approx -\Delta\rho^2/(2F^2)$.

5.10 Laser rangefinder emits pulses with pulse duration $\tau = 1$ ns. Calculate the period T between two pulses that would allow to measure the target velocity of 100 km/h with error not exceeding 1%.

Solution (5.10). The single pulse error ΔL equals

$$\Delta L = 0.5\tau c = 0.5 \cdot 1 \cdot 10^{-9} \cdot 3 \cdot 10^{8} = 0.15 \text{ m}.$$

To measure the velocity we need to do two measurements, separated by period T. Since the errors are statistically independent, the total error for two measurements would be

$$\Delta L_t = \sqrt{2(\Delta L)^2} \approx 0.21 \text{ m}.$$

To keep the error below 1%, the total travel distance between two measurements should be 100 times the error: $L = \Delta L_t/0.01 = 21$ m. Since 100 km/h equals approximately $V \approx 28$ m/s, the period T should be larger than

$$T > \frac{L}{V} \approx 0.75 \text{ s}.$$

5.11 Calculate the base B of a triangulation rangefinder to achieve $\Delta L = 10$ m at a 1 km distance. Assume that in both arms an afocal system with magnification $M = 10$ is used. Assume the eye can resolve non-coincident images separated by $\Delta\alpha = 0.5'$ of arc.

Solution (5.11). Using (5.19) we have

$$B \approx -\frac{L^2}{\Delta L}\frac{\Delta\alpha}{M} = \frac{1000 \cdot 1000 \cdot 0.000145}{10 \cdot 10} = 0.145\text{m}.$$

5.12 The rangefinder of a Leica film camera has base $B = 68$ mm and magnification $M = 0.72$. Estimate the focusing error at a distance of $L = 2$ m. Assume the eye can resolve non-coincident images separated by $\Delta\alpha = 0.5'$ of arc.

Solution (5.12). Using (5.19) we have

$$\Delta L \approx -\frac{L^2}{B}\frac{\Delta\alpha}{M} = \frac{2 \cdot 2 \cdot 0.000145}{0.068 \cdot 0.72} = 0.012 \text{ m}.$$

5.13 Design Cassegrain analog of Hubble space telescope with parameters: primary diameter $D_1 = 2.4$ m, mirror spacing $d = 5$ m, distance from secondary to focus $S_2' = 6$ m, equivalent focal length $F_{eq} = 57$ m. Determine, R_1, R_2, and the conic constants for the primary and secondary.

Solution (5.13).

Using (5.1) we find

$$F_1' = \frac{dF_{eq}'}{F_{eq}' - S_2'} = 5.588$$

$$F_2' = \frac{S_2' d}{S_2' - F_{eq}' + d} = -0.652.$$

Finally $R_1 = 2F_1' = 11.176$ concave, $R_2 = 2F_2' = -1.304$ convex. $S_2 = d - F_1' = -0.588$ The conic constant, according to (4.14):

$$e^2 = -\left[\frac{S_2 - S_2'}{S_2 + S_2'}\right]^2 \approx -1.482$$

Zemax simulation of the system with calculated parameters, illustrated by Fig. 5.17 demonstrates close to diffraction-limited performance on-axis in a field of view of $\pm 0.04°$.

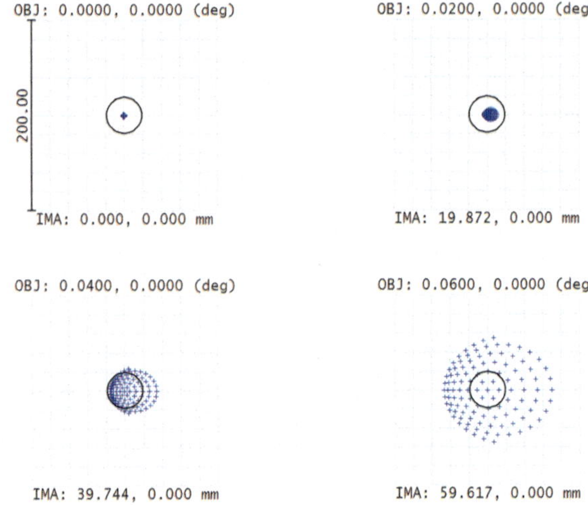

Fig. 5.17 To Problem 5.13: spot diagrams over the field for Cassegrain analog of Hubble telescope. Black circle corresponds to the diffraction limit of the spot size at $\lambda = 650$ nm

References

1. D.H. Hubel, *Eye, Brain, and Vision*. Scientific American Library Series. (Scientific American Library, 1988). ISBN: 9780716750208
2. Zemax Corporation. *Zemax Optical Studio*. http://www.zemax.com
3. J. Southall (ed.), *Helholtz's Treatize on Physiological Optics* (OSA, 1924)
4. Luizov, *Eye and color (in Russian)* (1983)
5. Kravkov, *Eye and Its Work* (1950)

Chapter 6
Applications of Coherent Light

Abstract Coherence, defined by the observability of interference, is fundamentally related to basic optical concepts such as the wavefront, point source, and extended object. Coherence theory provides the correct physical description of light emitted by polychromatic, extended objects. In practical terms, it defines the conditions under which such complex natural sources can be treated as if they were simple, point-like, monochromatic sources. Understanding coherence is critical for several reasons: • The temporal coherence of light imposes fundamental limitations on the design of interferometers for metrology. • The spatial coherence of light is defined by the angular size of an incoherent reference source. • The spectral properties, noise, and resolution of images are dependent on the coherence of the illumination. In general, imaging with coherent light provides lower resolution and higher noise due to speckle and interference artifacts. • Coherent light is used in holografy for engineering of coherent optical wavefronts.

6.1 Interference and Coherence

The coherence of light implies the ability to produce observable intensity interference patterns. "Observable" may sound somewhat redundant, since the interference of any two or more electromagnetic waves always produces a modulated interference pattern. However, this modulation is not always observable. Light waves have very high frequency, which cannot be directly registered by the human eye or by modern detectors. Only the time-averaged intensity can be measured. Interference patterns can be observed only if a stable phase relation (coherence) is preserved during the observation time.

To illustrate, consider interference of two waves with amplitudes A_1 and A_2 and frequencies ω_1 and ω_2. Without loss of generality, the phase of the first wave can be considered fixed and taken as a reference, while the phase of the second wave includes two terms:

G. Vdovin, *Elementary Technical Optics*, UNITEXT for Physics,
https://doi.org/10.1007/978-3-032-08626-6_6

- A very fast, randomly changing phase of the light field. This phase is not directly observable.
- A constant, or very slowly changing, observable phase delay ϕ that depends on the geometry of the problem.

$$U_1 = A_1 \mathrm{e}^{-\mathrm{i}\omega_1 t}$$
$$U_2 = A_2 \mathrm{e}^{\mathrm{i}(-\omega_2 t + \phi)}.$$

The intensity of the combined field is given by

$$I = (U_1 + U_2)(U_1 + U_2)^* = U_1 U_1^* + U_2 U_2^* + U_1 U_2^* + U_2 U_1^*$$
$$I = A_1^2 + A_2^2 + 2A_1 A_2 \cos(t(\omega_1 - \omega_2) + \phi)$$
$$I = I_1 + I_2 + 2\sqrt{I_1 I_2}\left|\gamma_{1,2}\right| \cos(\phi) \tag{6.1}$$

The first two terms in (6.1) represent the individual wave intensities. The third term with

$$\gamma_{1,2} = \frac{\overline{U_1 U_2^*}}{\sqrt{I_1 I_2}} \tag{6.2}$$

describes the interference result. The value $\gamma_{1,2}$, also called the "complex degree of coherence," has the physical meaning of a normalized correlation coefficient between the oscillations of the two wave fields. In general, it can be registered in two separate points, either in time or in space. The real part of $\gamma_{1,2}$ defines the observability of interference effects in an optical experiment. The difference $\omega_2 - \omega_1$ determines the temporal modulation of the interference pattern. The phase $\varphi(t)$, averaged over the observation time, determines the visible contrast of the pattern. The phase ϕ defines the static intensity at the interference point if $\omega_1 = \omega_2$ and $\left|\gamma_{1,2}\right| > 0$.

Example (*Two wavelengths*) Consider interference produced by two closely spaced wavelengths of an Nd:YAG laser $\lambda = 1.064\,\mu\text{m}$ and an Nd:YLF laser $\lambda = 1.047\,\mu\text{m}$. The modulation frequency is given by

$$\omega_{min} = \omega_1 - \omega_2 = c\,|k_1 - k_2| \approx 3 \cdot 10^8 \left| \frac{2\pi}{1.064 \cdot 10^{-6}} - \frac{2\pi}{1.047 \cdot 10^{-6}} \right| = 2.87 \cdot 10^{13}$$

The interference pattern is modulated with frequency $2.87 \cdot 10^{13}$ Hz, which is far too high to be registered with any existing detector. This example illustrates that it is nearly impossible to observe interference between wave fields emitted by two independent light sources, even if they have almost identical wavelengths. However, interference of a wave emitted by a single source with itself is possible and has many practical applications in metrology and imaging.

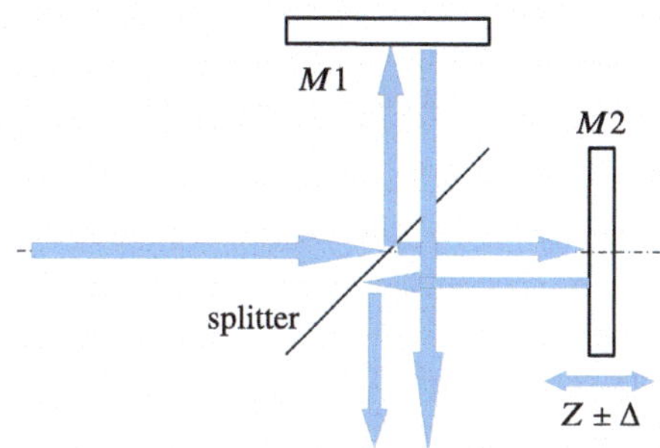

Fig. 6.1 Michelson interferometer for demonstration of interference of a light wave with a delayed copy of itself. The interference depends on the path difference between the waves reflected from mirrors M1 and M2

Example Consider the Michelson interferometer formed by a beamsplitter and two mirrors M1 and M2, as shown in Fig. 6.1.

The phase of the beam reflected from M2 is delayed with respect to the beam reflected from M1 by $\phi = k\Delta$, where $k = 2\pi/\lambda$. In this configuration the dynamic interference between two waves with different frequencies can be easily observed by moving one of the interferometer mirrors. Assume the mirror M2 is moved along the optical axis with velocity V. Then the reflected wave experiences a Doppler shift

$$\delta\omega = kc(-2V/c) = -4\pi V/\lambda \tag{6.3}$$

and the intensity in the observation plane is modulated with frequency $\delta\omega$. The same result can be obtained without invoking the Doppler effect, by considering the phase shift caused by linear movement of mirror M2:

$$\delta\omega = \delta\varphi_2 = -2kV = -4\pi V/\lambda. \tag{6.4}$$

Now consider the interference term of (6.1), assuming $\omega_1 = \omega_2$:

$$I_{interf} = 2\sqrt{I_1 I_2}\,|\gamma_{1,2}|\cos(\phi). \tag{6.5}$$

The interference term depends on the phase difference between the waves. If both phases φ and ϕ remain constant over time, the interference term has a fixed value, producing constant intensity in the range from 0 to $I_1 + I_2 + 2\sqrt{I_1 I_2}$.

However, if the phase φ, measured at different times, is uniformly distributed between 0 and 2π, the average interference term becomes

$$I_{interf} = \frac{2\sqrt{I_1 I_2}}{2\pi}\int_0^{2\pi}\cos(\varphi)d\varphi = 0. \tag{6.6}$$

In this case the observable result is a simple sum of the intensities, independent of the phase relations between the interfering waves.

In practice, the observability of interference depends on several factors:

- The frequencies of the interfering waves. Interference is observable if the frequencies are exactly equal ($\omega_1 = \omega_2$), since then the intensity is constant in time and depends only on the phase difference. Interference is also observable as temporal modulation if the beat frequency $|\omega_1 - \omega_2|$ is low enough to be detected.
- The spectral properties of the interfering waves. A purely monochromatic wave always produces an observable interference pattern with a delayed copy of itself, for instance in a Michelson interferometer. For a wave with broad spectrum, the visibility of interference depends on the temporal delay, which can also be expressed as the geometric path difference between the wave and its copy. The spectral properties of the source that influence its ability to interfere with itself define its temporal coherence.
- The angular size of an extended source when observed from the interference point. A two-slit interferometer produces clear fringes when illuminated by a monochromatic point source. However, the visibility of fringes decreases with the angular size of the source, even if it remains monochromatic. The spatial properties of the source that determine its ability to produce visible interference fringes define its spatial coherence.

6.1.1 Temporal Coherence

Consider two monochromatic waves $Ae^{i\omega_0 t}$ and $Ae^{i\omega_0(t+\Delta t)}$. The first wave is switched "on" at the moment $t = 0$ and then switched "off" at the moment τ. The second wave is switched "on" at the moment $t + \Delta t$ and then switched "off" at the moment $t + \Delta t + \tau$. Obviously, interference is possible only if $\Delta t < \tau$, because otherwise the waves will arrive at the interference point at different times.

If the wave is formed by a number of wave packets, each having duration τ, then interference of a wave packet with a delayed copy of itself is possible only if the delay is smaller than τ. For a larger delay, all waves will have a random phase with respect to the original state, resulting in zero averaged interference. In this situation τ is called the "coherence time."

For a wave propagating with speed c, the coherence length can be derived from the coherence time as

$$l_c = c\tau. \tag{6.7}$$

The distance $c\tau$ has the physical meaning of the maximum optical path difference between the arms in the interferometer shown in Fig. 6.1, still producing visible interference.

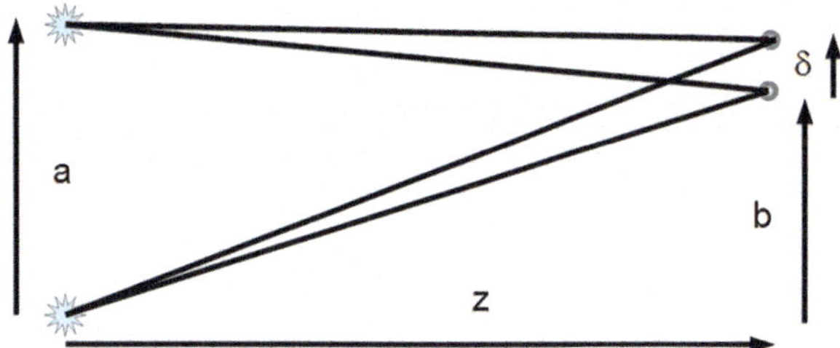

Fig. 6.2 Illustration of interference of light waves emitted by two incoherent point sources

To establish the relation between the coherence time and the spectrum of the optical signal, consider the Fourier transform of a wave packet with frequency ω_0 and duration τ:

$$U(\omega) = \frac{1}{\tau}\int_{-\tau/2}^{\tau/2} A\mathrm{e}^{\mathrm{i}\omega_0 t}\mathrm{e}^{-\mathrm{i}\omega t}dt = \\ = A\frac{\sin\big(0.5(\omega-\omega_0)\tau\big)}{0.5(\omega-\omega_0)\tau}. \tag{6.8}$$

This expression can be rewritten using the well-known $\mathrm{sinc}(x) = \sin(x)/x$ function:

$$A^2\frac{\big[\sin\big(0.5(\omega-\omega_0)\tau\big)\big]^2}{\big[0.5(\omega-\omega_0)\tau\big]^2} = \\ = A^2\big[\mathrm{sinc}(0.5(\omega-\omega_0)\tau)\big]^2 = A^2\big[\mathrm{sinc}(\pi(\nu-\nu_0)\tau)\big]^2. \tag{6.9}$$

This function has zeros at points defined by the solutions of the equation $\pi(\nu - \nu_0)\tau = n\pi$, with the first zero corresponding to

$$\Delta\nu = \nu - \nu_0 = \frac{1}{\tau}. \tag{6.10}$$

Now we can establish the relation between the coherence length and the bandwidth of the source expressed in terms of its wavelength:

$$l = c\tau = \frac{c}{\nu-\nu_0} = \frac{c}{c/\lambda - c/\lambda_0} = \left.\frac{\lambda^2}{\Delta\lambda}\right|_{\lambda_0\approx\lambda}. \tag{6.11}$$

Example: A LED with a wavelength of $\lambda_0 = 650\,\mathrm{nm}$ has a bandwidth of $2\Delta\lambda = 10\,\mathrm{nm}$. Determine the requirements for alignment of the Michelson interferometer shown in Fig. 6.1 that assure good fringe visibility.

Solution. For good fringe visibility, the maximum misalignment Δ between the lengths of the two arms should not exceed half of the coherence length of the source. The factor of one-half is taken because the path is doubled after reflection from the mirror. According to (6.11) we obtain:

$$\Delta \ll \frac{\lambda^2}{2\Delta\lambda} = \frac{650^2}{10} = 42250 \text{ nm} \sim 40 \text{ µm}.$$

If the source has bandwidth $\Delta\nu$, then to calculate the coherence length we recall that $\lambda = c/\nu$, where c is the speed of light. Then $\Delta\lambda = -\Delta\nu \cdot c/\nu^2$, and

$$l = \frac{c}{\Delta\nu} = c\tau.$$

To produce observable interference, the path difference between the interfering beams should be smaller than the coherence length.

6.1.2 Spatial Coherence

Spatial coherence is defined between two points in space. By definition, all points that belong to a spherical wavefront emitted by a point source are coherent. Indeed, since the source is a single point, it emits a spherical wave into a solid angle of 4π [sr]. All points at a distance R belong to the same wave packet and have the same phase delay. Therefore, they are coherent.

When the source is represented by a number of incoherent point sources, the degree of coherence becomes dependent on the geometry of the problem.

Consider the simplest case of two mutually incoherent point sources of equal intensity, separated by a distance a, with wavelength λ and wave number $k = 2\pi/\lambda$, as shown in Fig. 6.2. Our goal is to find how the degree of coherence between two points at distance z from the source depends on the transversal distance δ between the observation points. Thus, the coherence function depends on two coordinates: the separation between sources a, which defines the size of the source, and the separation between the observation points δ.

Assume the interference point at distance z has a transversal coordinate b. The optical paths to the point of interference are given by:

$$L_1 = \sqrt{z^2 + b^2}$$
$$L_2 = \sqrt{z^2 + (a - b)^2}.$$

Assuming $z \gg a$ and $z \gg b$, the maximum possible phase difference φ between the point located at b and its neighbor at $b + \delta$ is

$$\varphi \approx k\frac{d(L_1 - L_2)}{db}\delta \approx k\frac{a\delta}{z}. \tag{6.12}$$

For the field to be coherent, the phase difference within the patch δ should be limited: $|\varphi| < \pi$. Substituting this into (6.12) gives the size of the patch δ at which the oscillations remain coherent:

$$\delta < \frac{\pi z}{ak} = 1.22\frac{\lambda}{2A}, \tag{6.13}$$

where $A = z/(2a)$ is the numerical aperture (half the angular diameter) of the incoherent source, and $2A$ is the full angular diameter of the source.

Adding any intermediate incoherent source along the line between the two original sources does not reduce the degree of coherence. We can therefore generalize that the size of a coherent patch δ, positioned at distance z from an incoherent source of size a, is approximately defined by expression (6.13) and matches the one for diffraction-limited resolution of optical system. This formula represents the essence of the van Cittert–Zernike theorem. In imaging terms, the coherence patch has the size of the imaging aperture that is just sufficient to image an incoherent object of size a as a point source.

For a circular source, expression 6.13 takes the form

$$\delta < \frac{1.22\pi z}{ak} = 0.61\frac{\lambda}{A}. \tag{6.14}$$

An interferometer made from two pinholes, illuminated by an extended monochromatic source, will produce observable fringes if the distance between the pinholes is smaller than the size of the coherence patch δ.

Example. The angular diameter of the Sun is $\beta \approx 2000$ arc seconds. Define the size of the coherence patch.

Solution. Expression (6.13) can be rewritten as $\delta < 1.22\lambda/\beta$. Assuming $\lambda = 550$ nm and $\beta \approx 0.01$ rad, we obtain

$$\delta < \frac{550 \times 10^{-9}}{0.01} = 6.7 \times 10^{-5}\ \text{m} = 67\ \mu\text{m}.$$

Thus, solar light is spatially coherent within a patch of about 67 μm. A lens with a diameter smaller than this coherence patch would not resolve any details of the Sun image, producing a point-like stellar image defined purely by diffraction.

6.2 Coherent and Incoherent Imaging

The coherence of light directly influences the properties of an imaging system. Images of the same object, formed with spatially coherent illumination, can be drastically different from those formed with incoherent light. Coherent images contain observable interference terms that produce artifacts not present in the original. In some cases, the image of the object under coherent illumination looks quite different from the "real" view, as illustrated in Fig. 6.3. Here, the incoherent image, though blurred, correctly represents the object structure, while the coherent image looks sharper due to high-frequency, high-contrast artifacts that are not present in the original.

Figure 6.4 illustrates the difference between coherent (top) and incoherent (bottom) illumination of an object in an imaging system with numerical aperture A. In the coherent case, the illumination is homocentric: each feature of the sample is illuminated from a single direction. The scattered light fills the aperture of the imaging system. Then, the maximum resolved spatial frequency is defined as λ/A, where A is the NA of the imaging system.

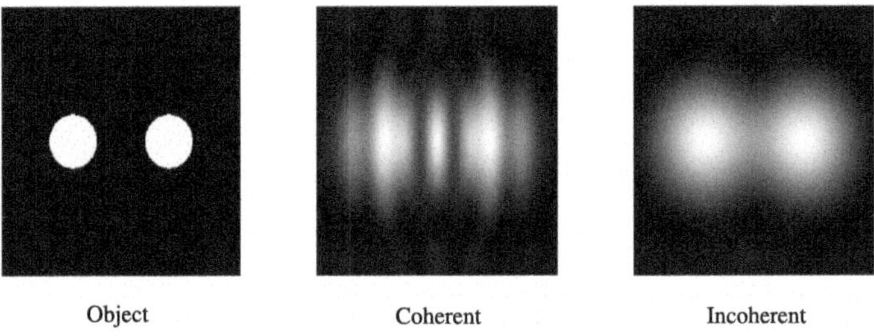

Fig. 6.3 Illustration of coherent and incoherent imaging of two bright spots

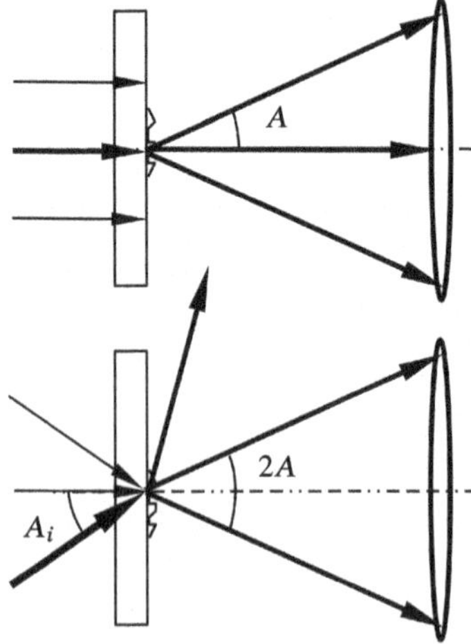

Fig. 6.4 Coherent (top) versus incoherent (bottom) illumination

In the incoherent case, each point of the object is illuminated with rays incident in the range $\pm A_i$, where A_i is the NA of the illumination system. If the illumination includes the marginal ray, $A_i = A$, meaning the NA of the illumination equals that of the imaging, then the total resolution of the system will be defined by $\lambda/(2A)$. In this case, the maximum spatial frequency is determined by the total scattering angle of $2A$, subtended by the aperture.

To realize incoherent illumination, the numerical aperture of the illumination system should be equal to the numerical aperture of the imaging lens. In general, the maximum resolved spatial frequency is given by

$$F_{max} \sim \frac{\lambda}{A + A_i}, \tag{6.15}$$

where A is the NA of the imaging system, and A_i is the NA of the illumination system (condenser).

The condition 6.15 means that the coherence patch created by the illumination system should be equal to the minimum size resolved by the lens. When $A_i < A$, the resulting images may appear visually better, as the micro-contrast is enhanced by interference effects and artifacts, as illustrated in Fig. 6.3. However, such apparent enhancement comes at the cost of significant loss of authentic information about the fine structure of the sample.

In microscopy, to achieve the maximum possible resolution, the illumination is formed by condensers described in Sect. 5.5. In the ideal case, the NA of the condenser should be equal to the NA of the imaging objective.

Coherent imaging also suffers from speckle noise. Optical speckle is a granular intensity distribution that appears when coherent light is scattered by a random screen. Each scattering element contributes a wave with a random phase, and the interference of many such contributions produces a random, fine-scale structure of bright and dark spots. Although the pattern appears chaotic, its statistical properties are well defined and reproducible.

The average speckle size is primarily determined by the wavelength and the numerical aperture of the scatterer. In the far field, the speckle grain size d is given by

$$d \approx \frac{\lambda}{D} L \tag{6.16}$$

where λ is the wavelength, L the distance to the observation plane, and D the aperture. The speckle intensity I has negative exponential probability density:

$$P(I) = \frac{1}{\langle I \rangle} \mathrm{e}^{\left(-\frac{I}{\langle I \rangle}\right)}, \tag{6.17}$$

where $\langle I \rangle$ is the mean intensity. This distribution defines high contrast of grained speckle structure.

In imaging, speckle reduces both the signal to noise ratio and image resolution. On the practical side, analysis of speckle patterns is used in metrology for sensing of surface displacement, flow visualization, and biomedical diagnostics. Recent advances include speckle correlation imaging and wavefront shaping through scattering media. Detailed description of properties of speckle fields can be found in [1].

A very detailed and fundamental treatment of different aspects of coherence of light is given in [2].

6.3 Holography

Gabor [3] demonstrated that when two optical waves U_1 and U_2 interfere, and the resulting intensity pattern $(U_1 + U_2) \cdot (U_1 + U_2)^*$ is recorded, illumination of this pattern by one of the original waves leads to the reconstruction of the other. This can be illustrated with two plane waves propagating at an angle α to each other. Let the complex amplitudes of the waves be

$$\begin{aligned} U_1 &= 0.5, \\ U_2 &= 0.5 \cdot \mathrm{e}^{ik\alpha x}. \end{aligned}$$

The resulting intensity I in the plane of interference is

$$\begin{aligned} I = (U_1 + U_2) \cdot (U_1 + U_2)^* &= 0.5 + 0.25\mathrm{e}^{ik\alpha x} + 0.25\mathrm{e}^{-ik\alpha x} \\ &= 0.5\cos(k\alpha x) + 0.5. \end{aligned} \tag{6.18}$$

This expression describes a sinusoidal intensity grating with absorption oscillating between 0 and 1. The hologram records this distribution so that its transparency T replicates the intensity, $T = I$.

To reconstruct the wave U_1 by illuminating the hologram with wave U_2, we evaluate $U_2 \cdot T$:

$$\begin{aligned} U_2 T &= 0.5\mathrm{e}^{ik\alpha x}\left(0.5 + 0.25\mathrm{e}^{ik\alpha x} + 0.25\mathrm{e}^{-ik\alpha x}\right) \\ &= 0.25\mathrm{e}^{ik\alpha x} + 0.125\mathrm{e}^{2ik\alpha x} + 0.125 \\ &= 0.5U_2 + 0.25U_1 + 0.25U_2\mathrm{e}^{ik\alpha x}. \end{aligned}$$

The result contains three terms: the reconstructed wave U_1 with reduced amplitude, the original illuminating wave U_2, and a spurious wave propagating symmetrically to U_1. The processes of recording and reconstruction are illustrated in Fig. 6.3.

Since any wavefront can be decomposed into plane waves, holography enables the recording and reconstruction of arbitrary complex wavefronts. The visibility of the recorded hologram depends on the mutual coherence of the interfering beams;

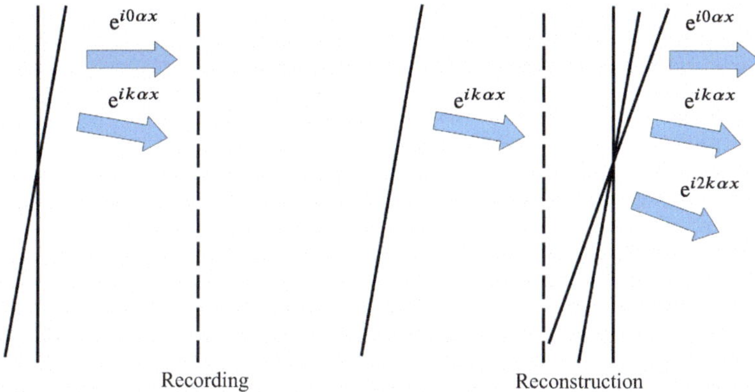

Fig. 6.5 Recording and reconstruction of a hologram. In the recording phase, the interference of two waves is recorded as a spatial grating. In the reconstruction phase, the grating is illuminated by one of the waves, reconstructing the original waves, and producing a spurious diffracted component

hence holographic setups require high temporal coherence and a mechanically stable environment.

Photographic emulsion, photo-polymers and nonlinear crystals are used for hologram recordings. In digital holography, the holograms are registered by electronic image sensors, with further reconstruction performed in a computer.

To produce a specific distribution of complex amplitude, the holographic pattern can be generated by a computer and fabricated using lithographic technology. Computer-generated holograms are used in optical shop testing to produce reference aspherical weavefronts for control of strong aspherical surfaces, in optical lithography to produce 3D lithographic patterns, and in laser beam processing to control and correct the intensity and phase distribution in a laser beam.

Holography is widely applied in metrology and nondestructive testing. Holographic interferometry allows precise measurement of displacements, vibrations, and stress distributions in mechanical structures. In microscopy, digital holography provides quantitative phase imaging of transparent specimens such as living cells, enabling label-free biomedical diagnostics. In data storage, multiplexed holograms achieve very high information density. For security and authentication, embossed holograms are used on credit cards and banknotes due to their resistance to forgery. Holography also finds use in laser beam shaping, fabrication of diffraction gratings, and other applications where control of the optical wavefront is essential. Its power lies in the ability to record and reconstruct the complete complex coherent wavefront, preserving both amplitude and phase.

6.4 Conclusion

The coherence of light determines the conditions under which interference phenomena can be observed and exploited in optical instruments. Temporal coherence is defined by the spectral width of the source and sets the maximum optical path difference allowable in interferometric systems. Light from narrowband lasers produces long coherence lengths, enabling measurements at large path difference lengths. Broadband sources such as LEDs restrict the usable path difference.

Spatial coherence is defined by the angular extent of the source as seen from the observation point. Point sources or sources limited by a small angular diameter provide high spatial coherence. The van Cittert–Zernike theorem relates the angular size of the source and observation geometry to the aperture over which light is coherent.

In imaging, coherence affects both resolution and signal to noise ratio. Coherent illumination enhances high-frequency contrast, but introduces interference artifacts and speckle. Incoherent illumination suppresses these artifacts and doubles the effective resolution limit compared to coherent imaging, provided the numerical apertures of illumination and imaging systems are matched. This principle defines the role of condensers in microscopy.

Interferometers such as Michelson or Fizeau rely directly on temporal coherence to measure optical path differences, surface deformations, and coherence properties. Spatial coherence governs fringe visibility in Young's two-slit arrangements, holography, and speckle metrology. Both temporal and spatial coherence thus impose fundamental design constraints on optical instruments.

Holography records both amplitude and phase of a coherent wave, enabling faithful reconstruction of optical wavefronts with applications in metrology, beam shaping, and imaging.

Altogether, the concept of coherence provides the framework for linking source properties to the performance of interferometers, microscopes, and imaging systems, ensuring that light from real, extended, and polychromatic sources can be correctly treated in optical engineering.

6.5 Problems

6.1 Red LED with wavelength of 630 nm and FWHM (full width half maximum) bandwidth of 20 nm is used as a light source in Michelson interferometer. Estimate the maximum difference δ between the lengths of the interferometer arms, that still allows to observe the fringe pattern.

Solution (6.1). *The difference δ equals half the coherence length, because the path difference doubles with the reflection on the shifted mirror. Then*

$$\delta = \frac{\lambda^2}{2\Delta\lambda} = \frac{630^2}{2 \cdot 20} \approx 10000 \text{ nm}$$

6.2 The Fizeau interferometer has $\delta = 2$ mm gap between the reference and test surfaces. You plan to use red LED with $\lambda = 650$ nm as the light source. Define the requirement to the bandwidth $\Delta\lambda$ of the LED.

Solution (6.2). *The coherence length should be larger than the double gap thickness δ: Then*

$$\Delta\lambda > \frac{\lambda^2}{2\delta} = \frac{650^2}{2 \cdot 2000000} \approx 0.1 \text{ nm}$$

6.3 Little girl is making a sunbeam using little square mirror with side A. Find the distance L at which the sunbeam projected on a wall has round shape. The angular size of Sun is $\alpha \approx 10^{-2}$.

Solution (6.3). *In the geometric approximation, single point of the mirror produces round image of Sun with diameter of αL. For the sunbeam to be round, the condition $\alpha L \gg A$ should be satisfied. $L \gg A/\alpha$. For solar image, the distance should be at least 100 times larger than A.*

6.4 Calculate the bandwidth of a laser, to be used in Mikhelson interferometer with the length of the first arm of $l_1 = 0.1$ m, and the second arm of $l_2 = 100$ m.

Solution (6.4). *The coherence length should satisfy the condition $l_c = c/\Delta\nu \gg 2|l_2 - l_1|$, then*

$$\Delta\nu \ll \frac{c}{|l_2 - l_1|} = 1.5 \text{ MHz}.$$

6.5 The Young interferometer is formed by two pinholes separated by $s = 1$ mm. At which distance L a monochromatic spatially incoherent light source with $\lambda = 500$ nm, and diameter of $D = 1$ cm, should be placed from the interferometer, to make fringes observable?

Solution (6.5). *According to van Cittert-Zernike theorem, the distance L should satisfy the condition*

$$\frac{1.22\lambda}{D} L > s,$$

then

$$L > \frac{sD}{1.22\lambda} \approx 16.4 \text{ m}.$$

6.6 The sun ray is reflected by a perfectly polished spherical metal ball with a diameter of $D = 4$ cm, and the reflected light illuminates the observer at a $L = 10$ m distance from the ball. Assuming the Sun has angular diameter of $2\alpha = 1900$ second of arc, estimate the radius of coherence Δ in the plane of observer. How this "reflected" spatial coherence is compared to the coherence of direct sunlight Δ_S? Assume wavelength of $\lambda = 550$ nm.

Solution (6.6). *The angular diameter of the Sun is* $\alpha \approx 10^{-2}$. *The scene is illuminated by the secondary image of the Sun, formed in the ball reflection. The diameter* $2h$ *of the secondary image of Sun is given by* $2h = F\alpha$, *where* F *is the focal length of the spherical ball surface, which is equal to a quarter of the ball diameter* D. *In our case* $F = 1\,cm$, *and* $2h = \alpha F = \alpha D/4$. *Then the angular size of the source is* $\varphi = 2h/L$, *and the radius of coherence* Δ *can be estimated as*

$$\Delta < 1.22\frac{\lambda}{\varphi} = 1.22\frac{4\lambda L}{\alpha D} = 0.067\text{ m}.$$

The coherence radius of direct sunlight is

$$\Delta_S < 1.22\frac{\lambda}{\alpha} = 67 \cdot 10^{-6}\text{ m}.$$

References

1. J.W. Goodman, *Speckle Phenomena in Optics: Theory and Applications* (Roberts & Company, 2007)
2. E. Wolf, *Introduction to the Theory of Coherence and Polarization of Light* (Cambridge University Press, 2007)
3. D. Gabor, Microscopy by reconstructed wave-fronts. Proc. R. Soc. Lond. Ser. A Math. Phys. Sci. **197**(1051), 454–487 (1949)

Chapter 7
Applications of Polarized Light

Abstract Polarization provides a powerful degree of freedom for controlling light in optical systems. A light wave can be represented as two orthogonal components, and their relative amplitudes and phases define its polarization state. Fully polarized light is described by Jones vectors and matrices. Partially polarized or unpolarized states require the Mueller formalism. This chapter introduces the mathematical frameworks for describing polarization, including Jones and Mueller representations, and outlines practical approaches to generating and analyzing linear and circular polarization. Common polarization elements such as polarizers, retarders, and birefringent prisms are discussed, together with devices for broadband control, such as the Fresnel rhomb. Applications range from microscopy and laser optics to liquid-crystal displays, optical isolators, and astronomy.

7.1 Jones (Maxwell) Vectors

Assume that the x and y components of polarized light are monochromatic and oscillate at the same frequency w. Then the wave is fully described by the amplitudes and phases of the two orthogonal components $E_x,\ E_y$ forming the Jones vector:

$$\begin{vmatrix} E_x \\ E_y \end{vmatrix} = \begin{vmatrix} A_x \exp(\mathrm{i}\varphi_x) \\ A_y \exp(\mathrm{i}\varphi_y) \end{vmatrix} \exp[\mathrm{i}(wt - kz)]. \tag{7.1}$$

For light linearly polarized in the y plane, $A_x = 0$.
For light linearly polarized in the x plane, $A_y = 0$.
For light linearly polarized at an angle of $\pi/4$ to the y axis, $A_x = A_y$ and $\varphi_x = \varphi_y$.
For right- or left-handed circular polarization, $A_x = A_y$ and $\varphi_x = \varphi_y \pm \pi/2$.

The total intensity of the polarized wave can be calculated as

$$I = |E_x^*, E_y^*| \begin{vmatrix} E_x \\ E_y \end{vmatrix} = A_x^2 + A_y^2. \tag{7.2}$$

G. Vdovin, *Elementary Technical Optics*, UNITEXT for Physics,
https://doi.org/10.1007/978-3-032-08626-6_7

If the light is described by the vector E_x, E_y at the input of an optical system, and by E'_x, E'_y at the output, then the transformation introduced by the system is described by the Jones matrix:

$$\begin{vmatrix} E'_x \\ E'_y \end{vmatrix} = \begin{vmatrix} J_{11} & J_{12} \\ J_{21} & J_{22} \end{vmatrix} \begin{vmatrix} E_x \\ E_y \end{vmatrix}, \tag{7.3}$$

where the four J-components of the Jones matrix are complex numbers defining the polarization conversion.

An ideal polarizer oriented in the x plane is described by the matrix

$$\begin{vmatrix} 1 & 0 \\ 0 & 0 \end{vmatrix}.$$

To obtain the matrix for an ideal polarizer rotated by an angle ϕ, we apply the rotation matrix twice: first to rotate the coordinate system into the polarizer plane, and then to return to the original coordinates:

$$\begin{vmatrix} \cos(-\phi) & \sin(-\phi) \\ -\sin(-\phi) & \cos(-\phi) \end{vmatrix} \begin{vmatrix} 1 & 0 \\ 0 & 0 \end{vmatrix} \begin{vmatrix} \cos(\phi) & \sin(\phi) \\ -\sin(\phi) & \cos(\phi) \end{vmatrix} = \begin{vmatrix} \cos^2(\phi) & \sin(\phi)\cos(\phi) \\ \sin(\phi)\cos(\phi) & \sin^2(\phi) \end{vmatrix}. \tag{7.4}$$

Expression 7.4 describes a linear polarizer in any orientation.

An ideal retarder introducing a phase delay φ into the x component is described by

$$\begin{vmatrix} \exp(i\varphi) & 0 \\ 0 & 1 \end{vmatrix}.$$

If light passes sequentially through a number of systems described by Jones matrices $|J_1|$, $|J_2| \dots |J_n|$, then the resulting transformation is given by

$$|E'| = |J_n| \dots |J_2||J_1||E|. \tag{7.5}$$

The parameters of the Jones matrix of a system can be determined by measuring the polarization state of the incoming and transmitted light. Since most optical measurements are based on intensity registration, the characterization of the Jones matrix requires a series of measurements with intensity detected through different linear and circular polarizers.

Finally, the Jones vector describes only the state of fully polarized light. Most natural light, however, is unpolarized or partially polarized. Such states cannot be described by the Jones vector alone.

7.2 Stokes Parameters and Mueller Matrices

The Jones formalism can be used only for fully polarized light. Practically, this means that the components $A_{x,y}$ and $\varphi_{x,y}$ must remain constant. If the polarization state changes with time, for example when the light has low temporal coherence and each wave packet has a different polarization, a more general formalism is required that takes into account time-averaged values.

In such general cases, any polarization state, including unpolarized and partially polarized light, is described by Stokes vectors and Mueller matrices. The Stokes vector is composed of four real values with the dimension of power density [W/m^2]:

$$\begin{vmatrix} S_0 \\ S_1 \\ S_2 \\ S_3 \end{vmatrix} = \begin{vmatrix} \langle A_x^2 \rangle + \langle A_y^2 \rangle \\ \langle A_x^2 \rangle - \langle A_y^2 \rangle \\ 2\langle A_x A_y \cos(\varphi_y - \varphi_x) \rangle \\ 2\langle A_x A_y \sin(\varphi_y - \varphi_x) \rangle \end{vmatrix}, \tag{7.6}$$

where $\langle\rangle$ denotes time averaging: $\langle a \rangle = \int_0^T a(t)dt$.

Parameter S_0 describes the total intensity. Parameters S_1 and S_2 describe the linearly polarized components. S_1 corresponds to components oriented along the x and y axes. S_2 corresponds to components rotated by 45° relative to x and y. A positive sign of S_1 indicates a predominance of the x-polarized component, while a negative sign indicates a predominance of the y-polarized component. A positive sign of S_2 indicates a predominance of polarization along $+\pi/4$, while a negative sign indicates a predominance along $-\pi/4$. Finally, S_3 describes the circularly polarized components, with positive values corresponding to right-handed and negative values to left-handed circular polarization.

For fully polarized light,

$$S_0^2 = S_1^2 + S_2^2 + S_3^2.$$

Equation 7.7 defines a spherical surface (*Poincaré sphere*) in the coordinate system defined by $S_{1,2,3}$. Any pure polarization state corresponds to a point on the surface of the Poincaré sphere.

For partially polarized light,

$$S_0^2 > S_1^2 + S_2^2 + S_3^2, \tag{7.7}$$

and the Stokes vector can be decomposed into two parts:

$$\begin{vmatrix} S_0 \\ S_1 \\ S_2 \\ S_3 \end{vmatrix} = \begin{vmatrix} S_0 - \sqrt{S_1^2 + S_2^2 + S_3^2} \\ 0 \\ 0 \\ 0 \end{vmatrix} + \begin{vmatrix} \sqrt{S_1^2 + S_2^2 + S_3^2} \\ S_1 \\ S_2 \\ S_3 \end{vmatrix},$$

where the first term represents the fully unpolarized part, and the second the fully polarized part. The degree of polarization is given by:

$$\Pi = \frac{\sqrt{S_1^2 + S_2^2 + S_3^2}}{S_0}.$$

The components of the Stokes vector can be determined experimentally by measuring the total intensity I_0, the intensities after passing through horizontal I_x and vertical I_y polarizers, the intensities after polarizers oriented at $\pi/4$ ($I_{\pi/4}$) and $-\pi/4$ ($I_{-\pi/4}$), and the intensities of right- and left-handed circularly polarized light, I_r and I_l:

$$\begin{vmatrix} S_0 \\ S_1 \\ S_2 \\ S_3 \end{vmatrix} = \begin{vmatrix} I_0 \\ I_x - I_y \\ I_{\pi/4} - I_{-\pi/4} \\ I_r - I_l \end{vmatrix}. \tag{7.8}$$

An optical element that modifies the polarization state is described by a 4×4 Mueller matrix $|M|$:

$$|S'| = |M|\,|S|\,.$$

Representative Mueller matrices for common polarization elements are given in [1].

The Stokes parameters can be measured in practice by performing intensity measurements through a linear polarizer at different orientations, and through a linear polarizer combined with a quarter-wave plate. A detailed description of measurement techniques for Stokes parameters is given in [1].

7.3 Linear Polarization

In the simplest case, linearly polarized light can be obtained with a grid of parallel conductive wires, with the grid period comparable to the wavelength. Such a wire-grid polarizer acts as a metallic mirror for the optical field component polarized along the wires, and transmits the orthogonal component. Wire-grid polarizers are available for the far-IR and terahertz ranges. In the optical range they can be fabricated by patterning a fine metal grid on a transparent substrate using high-resolution lithography.

The Polaroid polarizer, patented in 1929 with numerous later improvements, is made of a thin polyvinyl alcohol polymer doped with iodine to provide conductivity. When stretched in one direction, the long conductive molecules align along the stretch axis. The polarizer then works similarly to a wire-grid device: it exhibits strong absorption for the polarization component parallel to the stretch direction, while transmitting the orthogonal component.

A high degree of polarization, though with considerable intensity loss, can be obtained by reflecting a light beam at the Brewster angle. At this incidence angle, the p-polarized component is fully transmitted, while the s-polarized component is partially reflected.

A reasonably high degree of polarization, without strong intensity loss, can be obtained by passing a light beam through a sequence of plane-parallel plates at Brewster incidence. The p-component, polarized in the plane of incidence, undergoes zero reflection, while the orthogonal s-component is partially reflected at each interface. The total intensity of the s-component is reduced as $I_n = I_0(1 - R_\perp)^N$, where N is the number of reflections and I_0 is the initial intensity of the s-component. In a similar way, Brewster windows inside laser resonators efficiently filter the s-component, resulting in strongly p-polarized laser output.

Very high degrees of polarization are obtained with birefringent prisms, such as Glan-Thompson, Glan-Taylor, and Glan-Foucault prisms. In these devices, a reflecting interface inside the birefringent material exhibits total internal reflection for one polarization state while transmitting the other. The transmitted beam is highly polarized, with some intensity loss, whereas the reflected beam contains a mixture of both polarization states.

7.4 Circularly Polarized Light

In circularly polarized light, the phase difference between the x and y components equals $\pm\pi/2$. The sign of the phase difference determines the handedness (direction of rotation) of the polarization. The definition of handedness can be confusing, as there are two conventions for right-handed polarization:

- Looking from the source: point the thumb along the direction of propagation; the curled fingers show the rotation direction.
- Looking toward the source: point the thumb toward the source; the curled fingers show the rotation direction.

In this text, we adopt the first convention: looking from the source, right-handed polarization corresponds to clockwise rotation of the electric field vector, with the y-component delayed by one quarter of a wavelength relative to the x-component.

Circular polarization can be produced by introducing a quarter-wave phase shift between the x and y components of linearly polarized light. This phase shift is achieved by passing linearly polarized light through a properly oriented quarter-wave plate, fabricated from birefringent crystalline material. The thickness T of the plate is given by:

$$T = \frac{(k + 0.25)\lambda}{n_s - n_f}, \tag{7.9}$$

where k is the order of the plate, and n_s and n_f are the slow and fast refractive indices. Zero-order waveplates ($k = 0$) provide the best optical performance but are

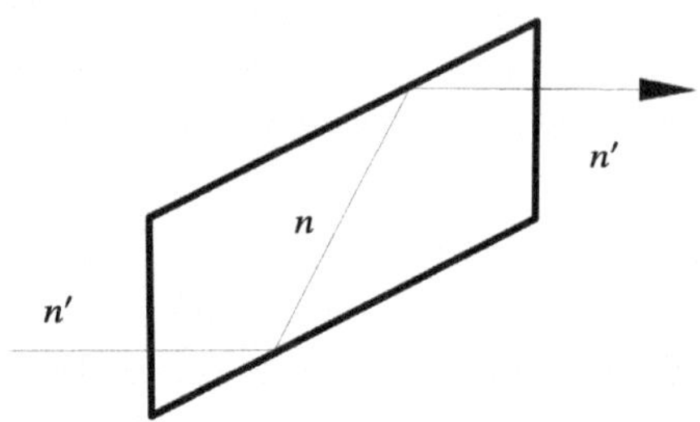

Fig. 7.1 Schematic of the Fresnel rhomb

the most difficult to fabricate. For crystalline quartz with $n_s = 1.556$ and $n_f = 1.547$ at $\lambda = 532$ nm, the zero-order thickness is $T = 14.78$ µm. Quarter-wave plates are fabricated by bonding a thin quartz layer to an isotropic glass substrate and polishing the quartz to the required thickness. Another approach combines two thicker, higher-order waveplates with crossed axes so that the integer-order retardance cancels, leaving only the residual quarter-wave delay.

As follows from Eq. 7.9, the performance of a quarter-wave plate is strongly wavelength dependent, especially at higher orders. The *Fresnel rhomb*, shown in Fig. 7.1, provides broadband conversion from linear to circular polarization by exploiting the $\pi/2$ phase shift between orthogonal polarizations after a sequence of total internal reflections in isotropic glass. According to [2], when light is reflected at a medium-to-air interface, the phase shift Δ between the two polarizations is given by

$$\tan\frac{\Delta}{2} = \frac{\sqrt{\sin^2\phi - N^2}\cos\phi}{\sin^2\phi}, \tag{7.10}$$

where $N = n'/n$ is the relative refractive index, and ϕ is the incidence angle. Substitution shows that to achieve a $\pi/4$ phase shift per reflection with $n = 1.5,\ n' = 1$, the incidence angle should be $\phi \approx 53°$. Two reflections inside the rhomb then provide the required $\pi/2$ total retardance. The phase delay of the Fresnel rhomb varies only slightly with wavelength due to material dispersion. For BK7 glass, the variation of Δ remains within 2% between 600 and 1550 nm, and does not exceed 5% at 400 nm.

The handedness of circularly polarized light can be reversed by introducing an additional π delay between the polarization components. This can be achieved either by:

- using a half-wave plate, a pair of properly oriented quarter-wave plates, or two Fresnel rhombs,
- reflection from a mirror at normal incidence.

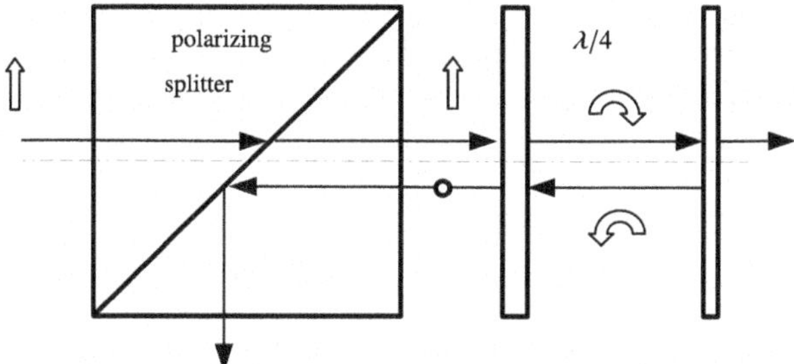

Fig. 7.2 Optical isolator. Linearly p-polarized light passes the polarization-sensitive splitter. The quarter-wave plate introduces a $\pi/2$ phase shift between the s- and p-components, first for the forward-propagating wave, and second, for the back-reflected wave, resulting in s-polarization. The s-polarized back reflection is then reflected by the polarizing beamsplitter and diverted away from the system

7.5 Applications of Polarized Light

In the more than 200 years since the first report of polarized light by the French physicist Biot, polarization has found numerous technical applications. Some examples are listed below, without detailed descriptions, since their principles can be understood using the basic concepts outlined in the previous sections:

- Lasers can produce highly coherent light with a very high degree of polarization.
- Optical isolators, shown in Fig. 7.2 are used to eliminate back reflection in optical systems.
- Observation of a transparent sample placed between crossed polarizers allows birefringent regions to be detected. These can be caused by internal stresses, impurities, or structural anisotropy. Polarized light microscopy is widely used to study crystals, minerals, and biological tissues.
- Polarized light is used in liquid crystal displays (LCDs) for TVs, monitors, and smartphones. The orientation of liquid crystal molecules modulates the polarization of transmitted light. This polarization modulation is converted into intensity modulation by a linear polarizer.
- Polarized light is used in 3D displays. Two images, filmed from viewpoints corresponding to the two eyes of the observer, are projected with orthogonal polarizations. Viewers wearing polarizing glasses perceive depth, since each eye receives only its corresponding image.
- The Faraday effect is widely used in high-power laser optics, particularly in amplifiers, to eliminate back-reflected beams. A "Faraday isolator" (or optical isolator), shown in Fig. 7.3, consists of an input linear polarizer that transmits the incoming linearly polarized light, and a Faraday cell that rotates the polarization by an angle $\beta = \nu BL$, where ν is the Verdet constant, B is the magnetic field collinear with the propagation direction, and L is the cell length. The forward-propagating

light experiences a 45° polarization rotation. Light reflected back through the cell experiences an additional 45° rotation, for a total of 90°. This rotated polarization is then blocked by the input polarizer or diverted by the polarizing splitter.

- Polarization densitometry relies on rotation of the plane of linear polarization. Linear polarization can be represented as the sum of two circularly polarized components: left-handed and right-handed. A class of materials exhibits *circular birefringence*, in which the propagation velocities of right- and left-handed components differ. Linearly polarized light passing through such *optically active* material remains linearly polarized, but its polarization plane rotates proportionally to the propagation distance. This effect should not be confused with circular polarization. Well-known optically active substances include sucrose (ordinary sugar), cholesterol, and chiral liquid crystals. Since the rotation of the polarization plane in sugar solution is proportional to concentration, this effect is used industrially to measure solution concentrations. The cuvette with solution is placed between crossed polarizers. The analyzer is rotated by an angle α to block transmission. For a fixed cuvette thickness, α is proportional to the concentration of sucrose solution.
- In open-air photography and surveillance, dielectric reflecting surfaces such as windows, wet roads, or water surfaces are illuminated by unpolarized sunlight at angles close to the Brewster angle, producing bright, strongly s-polarized glints. These glints saturate image sensors and obscure details behind reflections. Polarizing filters oriented appropriately can suppress the reflected s-polarized light, removing glare and improving image information. Since polarization can also be altered by dielectric mirrors, AR-coated surfaces, and other optics in the imaging system, linear polarizers are often combined with quarter-wave plates to produce circularly polarized light that is less sensitive to internal polarization effects inside the instrument.
- A non-zero degree of polarization in the unresolved image of a star can be evidence of an orbiting planet. Although the planet itself cannot be resolved, light reflected from it is polarized and contributes to the total polarization of the star's image. Periodic changes in the polarization degree can reveal the orbital period of the planet.

A detailed and fundamental treatment of polarization and its applications is given in [3].

7.6 Conclusion

The Jones formalism enables exact treatment of fully polarized light, with linear polarizers, retarders, and birefringent elements represented as matrices. Sequential action of elements is described by matrix multiplication, allowing direct modeling of cascaded systems. The Mueller–Stokes formalism extends this description to partially polarized and unpolarized fields, with measurable parameters providing

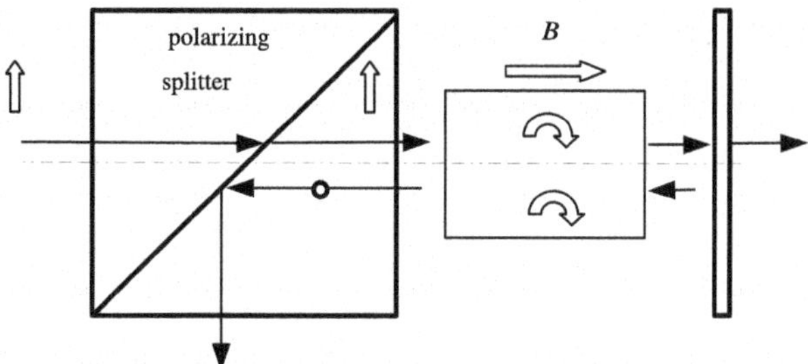

Fig. 7.3 Faraday isolator. The polarization plane is rotated in the Faraday cell by $\pi/4$ in both forward and backward directions. The reflected wave experiences a total rotation of $\pi/2$, causing it to be blocked (reflected away) by the polarizer

quantitative characterization of polarization states and their evolution through optical systems.

Physical control of polarization is realized through selective transmission, absorption, or phase retardation. Wire-grid and polymer polarizers provide linear polarization in broad spectral ranges. Brewster incidence allows separation of s- and p-components with moderate efficiency. Birefringent prisms achieve nearly complete separation. Retarders, including quarter- and half-wave plates, convert between linear, elliptical, and circular polarization. Fresnel rhombs provide broadband phase retardation with reduced sensitivity to wavelength.

Polarization control is widely used in display and communication systems. In material science, crossed polarizers reveal birefringent structures in crystals and tissues. In imaging and photography, polarizers suppress glare and glints from reflections on dielectric surfaces. In liquid crystal displays, polarization modulation translates molecular orientation into intensity modulation. In stereoscopic displays, orthogonal polarization channels encode separate images for the left and right eyes. High-power laser systems employ isolators based on Faraday rotation to suppress back-reflections. In astronomy, polarization measurements reveal scattering processes and enable indirect detection of exoplanets.

Polarization represents a functional degree of freedom in a broad range of scientific and engineering applications.

7.7 Problems

7.1 Can absolutely monchromatic light be randomly polarized? Assume ideal case with $\Delta\nu = 0$, where ν is the frequency of light.

Solution (7.1). *Absolutely monochromatic light can not be randomly polarized, because the polarization changes can occur in time frame that is longer than the coherence time, which is infinitely long for absolutely monochromatic light. However,*

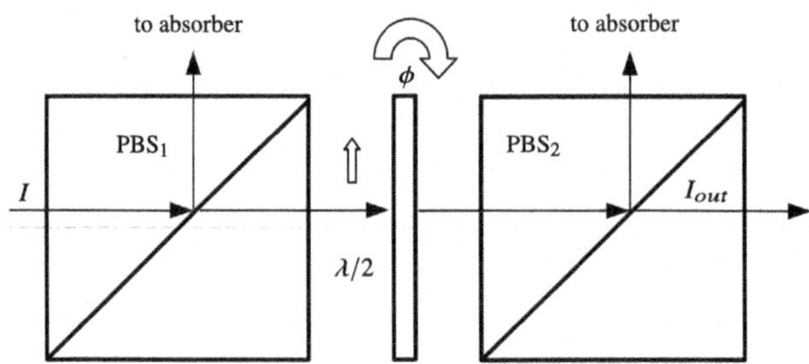

Fig. 7.4 Polarizing attenuator

laser light can be randomly polarized because the coherence time of any real laser is limited.

7.2 Light beam polarized in the horizontal plane propagates parallel to the surface of a horizontal table. Is it possible to rotate the beam polarization, making it polarized in the vertical plane, by using two metal-coated flat mirrors?

Solution (7.2). *The solution requires some out of the plane thinking.*

7.3 In Fig. 7.3 the Faraday cell rotates the polarization vector by angle ϕ in a single pass. Assuming 100% reflection on the flat mirror, calculate the leakage back as a function of ϕ, if the input beam is described by two components E_x and E_y at the entrance to the polarizing cube. The slow axis of the quarter-wave plate is oriented at 45° to the x axis.

Solution (7.3). *The wave is passing through cube, which can be described as a linear Y polarizer, then rotation of the coordinate system by the angle 2ϕ (double pass) by the Faraday cell, and linear Y polarizer for leakge back. The rejected beam is reflected down by the X arm of the polarizing cube. If the initial Jones vector of the plane-polarized wave is $\mathbf{E}_{in} = \begin{pmatrix} E_x \\ E_y \end{pmatrix}$, the resulting Jones vector $\mathbf{E}_{out}$ after passing through the system is:*

$$\mathbf{E}_{out} = \begin{pmatrix} 0 & 0 \\ 0 & 1 \end{pmatrix} \cdot \begin{pmatrix} \cos 2\phi & \sin 2\phi \\ -\sin 2\phi & \cos 2\phi \end{pmatrix} \cdot \begin{pmatrix} 0 & 0 \\ 0 & 1 \end{pmatrix} \cdot \begin{pmatrix} E_x \\ E_y \end{pmatrix}$$

The leakage intensity I_{out} is given by the squared magnitude of the resulting Jones vector $\mathbf{E}_{out}$:

$$I_{out} = |\mathbf{E}_{out}|^2 = |E_{x,out}|^2 + |E_{y,out}|^2 = E_y^2 \cos^2 2\phi,$$

then the rejected intensity, shown as reflected down in the splitter cube, is given by

$$E_y^2 \sin^2 2\phi,$$

and the total intensity is then $E_y^2(\cos^2 2\phi + \sin^2 2\phi) = E_y^2$. *Rotation of* $\phi = \pi/2$ *provides maximum rejection.*

7.4 Polarizing intensity attenuator consists of two polarizing beamsplitters and rotating half-wave plate as shown in Fig. 7.4. Calculate the intensity transmission I_{out}/I for randomly polarized input as function of plate rotation angle ϕ.

Solution (7.4). *If the initial Jones vector of the plane-polarized wave is* $\mathbf{E}_{in} = \begin{pmatrix} E_x \\ E_y \end{pmatrix}$, *the resulting Jones vector* $\mathbf{E}_{out}$ *after passing through the system is:*

$$\mathbf{E}_{out} = \begin{pmatrix} 0 & 0 \\ 0 & 1 \end{pmatrix} \cdot \begin{pmatrix} \cos\phi & \sin\phi \\ -\sin\phi & \cos\phi \end{pmatrix} \cdot \begin{pmatrix} 1 & 0 \\ 0 & -1 \end{pmatrix} \cdot \begin{pmatrix} \cos\phi & sin\phi \\ \sin\phi & \cos\phi \end{pmatrix} \cdot \begin{pmatrix} 0 & 0 \\ 0 & 1 \end{pmatrix} \cdot \begin{pmatrix} E_x \\ E_y \end{pmatrix} = \begin{pmatrix} 0 \\ -E_y \cos 2\phi \end{pmatrix},$$

and $I_{out} = E_y^2 \cos^2 2\phi$. *Then for* $I = 1$ *we can assume* $E_x = E_y = \sqrt{2}$, *and the output intensity oscillates between 0 and 0.5 as a function of* ϕ. *For polarized input with* $E_x = 0$ *and* $E_y = 1$, *the first polarizer can be omitted. In this case the output oscillates between 1 and zero as a function of* ϕ. *This scheme has important advantage: no power is dissipated in optical elements, the power is just re-distributed between the outputs of the second polarizing cube. The rejected beams, pointed up in the figure, can be directed to external absorber.*

7.5 The minimum thickness of the quartz plate, that can be reliably fabricated is $T = 0.25$ mm. Calculate the exact minimum thickness and the order of a quarter-wave plate fabricated with this technology. Assume crystalline quartz with slow $n_s = 1.556$, and fast $n_f = 1.547$ at $\lambda = 632$ nm.

Solution (7.5). *Using 7.9 we can calculate the order of interference as*

$$k = \frac{T(n_s - n_f)}{\lambda} - 0.25 = 3.31.$$

Then, assuming $k = 4$, *the exact thickness of the plate should be*

$$T = \frac{4.25(1.556 - 1.547)}{632 \cdot 10^{-9}} \approx 2.98 \cdot 10^{-3}\text{m}.$$

7.6 Design the quaterwave Fresnel rhomb for $\lambda = 550$ nm using (A) N-BK7 glass with glass number (MIL code) 517642, and (B) N-SF66 glass with glass number 923209. Calculate the phase error between the s and p components at 400 nm and 700 nm for both cases.

References

1. D.H. Goldstein, *Polarized Light* (CRC Press, 2017)

2. M. Born, E. Wolf, *Principles of Optics: Electromagnetic Theory of Propagation, Interference and Diffraction of Light*, 7th edn. (Cambridge University Press, 1999b). Note: A classic and comprehensive text on optics, including detailed discussions on polarization
3. E. Wolf, *Introduction to the Theory of Coherence and Polarization of Light* (Cambridge University Press, 2007)

Chapter 8
Optical Testing

Abstract This section describes the quality requirements for optical systems, encompassing specifications for aberrations, surface quality, material parameters, and roughness. It also introduces the most common methods for testing optical quality, such as stylus profilometry, interferometry, Hartmann-Shack sensing, and transport of intensity methods. Basic quality requirements for surface figure, roughness, alignment, and other parameters in optical systems comprising both mirrors and lenses are defined. Efficient metrology procedures to verify the implementation of these specifications are described.

8.1 Strehl Parameter

For diffraction-limited quality, a spherical wavefront emitted by a point in the object should be converted into a spherical wavefront converging to the corresponding image point. Any distortion of the wavefront sphericity reduces the image quality.

The Strehl parameter (St) is defined as

$$St = \frac{I_0}{I_a} \approx \mathrm{e}^{-\sigma^2}\Big|_{\sigma<1}, \tag{8.1}$$

where I_0 is the focal intensity for an ideal spherical wavefront, I_a is the actual focal intensity, and σ [rad] is the root mean square (rms) deviation of the wavefront from the reference sphere centered at the focus. Equation 8.1 is valid only for small aberrations $\sigma < 1$. For larger aberrations, the Strehl parameter depends on the detailed structure of the aberration. This is particularly important in the coherent case, when interference and speckle effects introduce random modulation and may reduce the axial intensity nearly to zero. However, for average (incoherent) estimates, a simplified engineering formula can be used with caution:

$$St = \frac{I_0}{I_a} \approx \frac{1}{1+\sigma^2}\Big|_{\sigma>1}. \tag{8.2}$$

G. Vdovin, *Elementary Technical Optics*, UNITEXT for Physics,
https://doi.org/10.1007/978-3-032-08626-6_8

For example, to achieve $St > 0.5$, the wavefront error must satisfy $\sigma < \sqrt{-\ln(0.5)} \approx$ 0.8 rad, which is approximately equivalent to $\lambda/8$.

In a system with N surfaces, assuming statistical independence of the aberrations introduced by surface imperfections $\sigma_1 \ldots \sigma_N$, the requirement can be written as

$$\sqrt{\sum_{i=1}^{N} \sigma_i^2} < \frac{\lambda}{8}. \tag{8.3}$$

If all surfaces contribute equally, the rms deviation per surface is $\sigma_i < \frac{1}{\sqrt{N}} \frac{\lambda}{8}$. The general requirement of $\lambda/8$ can be adjusted for specific cases: $\lambda/20$ corresponds to $St > 0.9$, and $\lambda/30$ corresponds to $St > 0.95$.

8.2 Surface Roughness

Mechanical roughness is characterized by measuring the surface profile over a defined sampling length. Roughness parameters are defined by ISO 4287, with the most important being:

- *Rz*: the maximum height of the profile within the sampling length.
- *Rq*: the root mean square deviation of the assessed profile.
- *Rdq*: the root mean square slope of the assessed profile.

Additional parameters and methods are defined in ISO 12085 and ISO 13565-(2,3).

The optical effect of surface roughness can be described by the total integrated scatter, which quantifies the fraction of scattered light as a function of wavefront rms deformation σ [1]:

$$T = T_0 \left[1 - \mathrm{e}^{-\left(\frac{\sigma \cos(\Theta)}{\lambda} \right)^2} \right], \tag{8.4}$$

where T_0 is the specular reflectivity or transmission of the surface without scattering, and Θ is the incidence angle. For reflective surfaces, $\sigma = 4\pi R_q$. For transparent transmitting surfaces, $\sigma = 2\pi(n-1)R_q$, where n is the refractive index. For the same R_q, transmitting optics scatters less than reflective optics. Therefore, mirrors must meet stricter R_q requirements than lenses. Equation 8.4 also shows that scattering depends on the incidence angle: surfaces reflecting at oblique incidence scatter less. X-ray telescope optics are designed to operate at very grazing incidence angles $\Theta \sim \pi/2$ to minimize scattering at very short wavelengths, where the atomic structure itself represents the surface roughness.

Commercially polished glass typically has R_q values between 1 and 5 nm, while superpolished surfaces can reach $R_q \approx 0.1$ to 0.2 nm. Micro-optical components fabricated using lithography, etching, or micromachining exhibit a wide range of R_q, depending on the fabrication technology.

8.3 Scratch/Dig Specification

The number and size of surface defects are important metrics of optical quality, specified by US standard MIL-PRF-13830B. The standard also specifies allowable bubble counts in optical glass. Scratches are characterized by a scratch number, with higher numbers corresponding to deeper scratches. Dig numbers correspond to defect diameters, expressed in units of 10 μm. Scratch/dig specifications apply separately to two zones: stricter tolerances are required in the inner zone near the optical axis, while tolerances are relaxed in the outer zone.

Relaxed specifications, such as 80/50 to 40/20, are used for large-aperture imaging optics such as telescopes and binoculars. Laser optics and mirrors for high-power lasers require much stricter specifications, typically 20/5 to 10/1. For more details and test procedures, refer to the text of the standard.

8.4 Stylus Profilers

Surface quality and figure can also be measured mechanically. For spherical surfaces, the sag s across a diameter D can be measured with a lens clock. The radius of curvature R and optical (focusing) power Φ are then calculated as (see Eq. 2.9 for the parabolic approximation of sag):

$$R = \frac{D^2}{8\,s}, \quad \Phi = \frac{n-1}{R}. \tag{8.5}$$

Lens clock dials are usually marked in diopters for a spherical air-glass interface with refractive index n, which is printed on the dial. The radius of curvature R can then be determined from the diopter reading D as

$$R = \frac{n-1}{D}. \tag{8.6}$$

Surface roughness can be measured by tracing the surface with a stylus, usually made of diamond with a spherical tip radius of several μm. The profile is recorded over several mm at sliding speeds of fractions of mm/s, depending on sensitivity and expected structure. Stylus measurements are suitable for evaluating surface microstructure and can also be used for figure control of micro-optical components such as holographic elements and microlens arrays. A comparison between stylus methods and optical profilometry is given in [2].

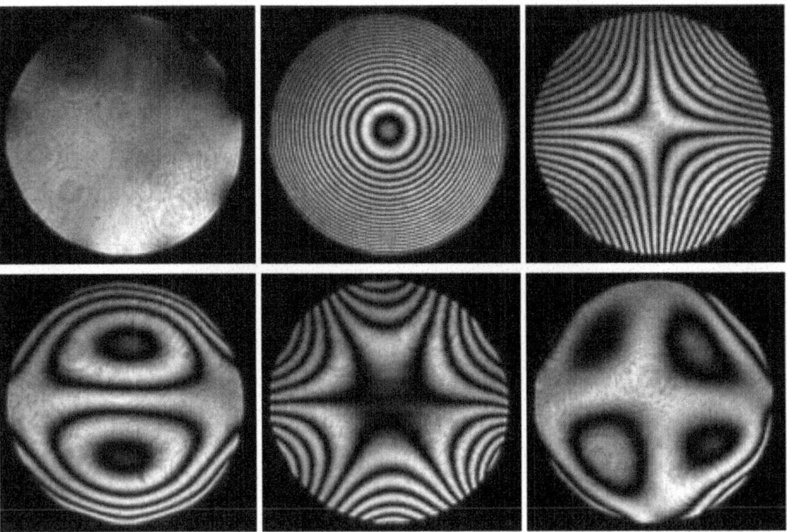

Fig. 8.1 Example interferograms obtained with a deformable mirror forming low-order aberrations

8.5 Control of Optical Surfaces

The optical figure of a surface can be measured directly with an interferometer, as shown in Fig. 8.1. Although very precise, interferometric testing is sensitive to vibrations and has limited dynamic range in surface sag and tilt, due to the maximum number and spatial frequency of resolved interference fringes. This is especially problematic when the test surface strongly deviates from the reference. In such cases, special holographic elements can be used to generate the required reference wavefront, matching the nominal surface shape so that only deviations appear as fringes.

The surface figure can also be reconstructed from surface tilts measured at multiple points. Hartmann and Shack–Hartmann sensors are based on this principle. These methods are less sensitive to external disturbances, but reconstruction assumes continuity of derivatives. Wavefront tilts can also be measured with Ronchi tests and shearing interferometers, both of which rely on self-interference of the wave with a modified copy of itself.

Finally, the figure can be reconstructed from the wavefront curvature, measured using the transport-of-intensity equation. These methods require highly sensitive linear intensity measurements, are susceptible to background light, and rely on assumptions about boundary conditions in solving the inverse problem.

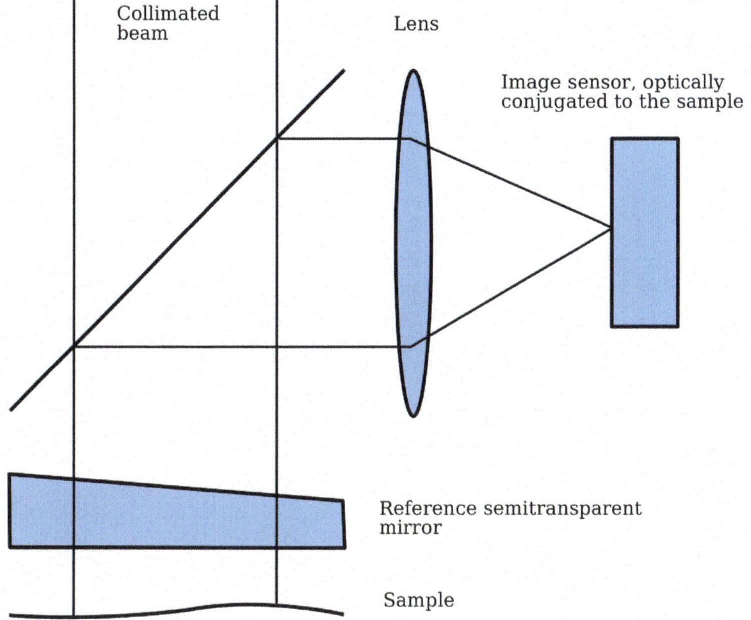

Fig. 8.2 Fizeau interferometer

8.5.1 Phase-Shift Interferometry

Two-wave interference is realized in the majority of measurement interferometers. The scheme of the Fizeau interferometer, widely used for testing of flat optics, is shown in Fig. 8.2. The intensity due to interference of two partially coherent waves, reflected by the reference and sample, is given by Eq. 6.1. This equation can be rewritten in a simplified form:

$$I = A(1 + B\cos(\varphi_0)), \quad B \leq 1, \tag{8.7}$$

where, of the three unknowns A, B, and φ_0, only the phase φ_0 is of interest, as it describes the difference between the reference and the measured wavefronts. Equation 8.7 can be trivially solved for $A = B = 1$ as $\varphi_0 = \pm\arccos(I - 1)$. However, this solution is ill-defined in the range $0 \ldots 2\pi$ and cannot be applied for arbitrary A and B.

A general solution can be obtained by measuring the intensity I for M equally spaced phase shifts. To achieve these shifts, the distance between the reference and the sample is changed with phase steps:

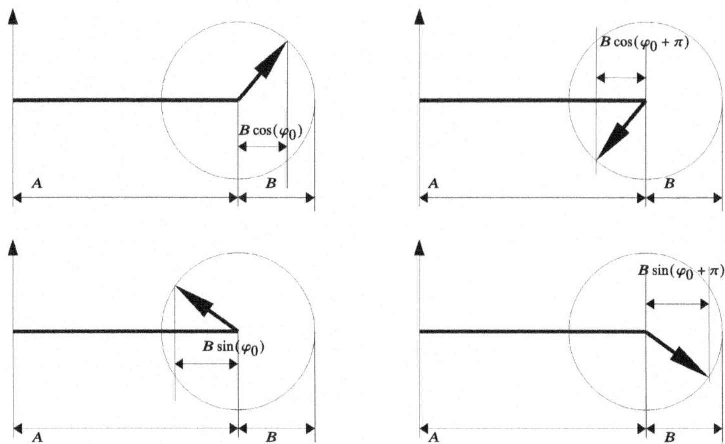

Fig. 8.3 Vector representation of the phase-shift method of phase reconstruction

$$\psi_n = 2\pi \frac{n}{M},$$
$$I_n = A(1 + B\cos(\varphi_0 + \psi_n)), \quad n = 0 \dots M-1. \tag{8.8}$$

In the simplest case, for $M = 4$, we obtain (see Fig. 8.3):

$$\begin{aligned} I_1 &= A(1 + B\cos(\varphi_0)), \\ I_2 &= A(1 + B\cos(\varphi_0 + \tfrac{\pi}{2})) = A(1 - B\sin(\varphi_0)), \\ I_3 &= A(1 + B\cos(\varphi_0 + \pi)) = A(1 - B\cos(\varphi_0)), \\ I_4 &= A(1 + B\cos(\varphi_0 + \tfrac{3\pi}{2})) = A(1 + B\sin(\varphi_0)). \end{aligned} \tag{8.9}$$

Combining terms yields a solution independent of A and B:

$$\begin{aligned} \frac{I_4 - I_2}{I_1 - I_3} &= \frac{\sin(\varphi_0)}{\cos(\varphi_0)}, \\ \varphi_0 &= \arctan\left[\frac{I_4 - I_2}{I_1 - I_3}\right]. \end{aligned} \tag{8.10}$$

In the general case, for $M \geq 3$, we calculate:

$$\begin{aligned} S_1 &= \sum I_n \cos(\psi_n) = \sum A(1 + B\cos(\varphi_0 + \psi_n))\cos(\psi_n), \\ S_2 &= \sum I_n \sin(\psi_n) = \sum A(1 + B\cos(\varphi_0 + \psi_n))\sin(\psi_n). \end{aligned}$$

Using $\cos(a + b) = \cos(a)\cos(b) - \sin(a)\sin(b)$, we obtain:

$$S_1 = A\sum\cos(\psi_n) + AB\left(\sum\cos(\varphi_0)\cos^2(\psi_n) - \sum\sin(\varphi_0)\cos(\psi_n)\sin(\psi_n)\right),$$
$$S_2 = A\sum\sin(\psi_n) + AB\left(\sum\cos(\varphi_0)\cos(\psi_n)\sin(\psi_n) - \sum\sin(\varphi_0)\sin^2(\psi_n)\right).$$

Since $\sum\cos(\psi_n) = \sum\sin(\psi_n) = \sum\sin(\psi_n)\cos(\psi_n) = 0$ and $\sum\cos^2(\psi_n) = \sum\sin^2(\psi_n) = M/2$, we obtain:

$$S_1 = AB\frac{M}{2}\cos(\varphi_0),$$
$$S_2 = -AB\frac{M}{2}\sin(\varphi_0).$$

Therefore,

$$\tan(\varphi_0) = -\frac{S_2}{S_1} = -\frac{\sum_0^{M-1} I_n\sin(\psi_n)}{\sum_0^{M-1} I_n\cos(\psi_n)}. \tag{8.11}$$

Equation 8.11 allows phase reconstruction with any number $M \geq 3$ of phase steps.

Phase-shift interferometry is the industrial standard, implemented in a number of calibrated instruments produced by Zygo, Wyko, etc.

8.5.2 *Fourier Phase Reconstruction*

The Fourier method, proposed in 1982 [3], reconstructs the phase profile from a single interferogram. The fitting interferogram must consist of almost parallel fringes without closed figures, such as rings. Such a pattern, example of which is shown in Fig. 8.4, can be obtained by introducing sufficient tilt into one arm of the interferometer. Let the fringe pattern have the form:

$$g(x, y) = c(x, y)\exp(i2\pi f_o x) + c^*(x, y)\exp(-i2\pi f_o x), \tag{8.12}$$

where

$$c(x, y) = \tfrac{1}{2}\exp(i\psi(x, y)), \tag{8.13}$$

$\psi(x, y)$ describes the wavefront, and $f_o x$ represents the tilt introduced in the setup.

The Fourier transform of Eq. (8.12) yields:

$$G(f, y) = A(f, y) + C(f + f_o, y) + C^*(f - f_o, y), \tag{8.14}$$

where A and C denote Fourier spectra, and f is the spatial frequency.

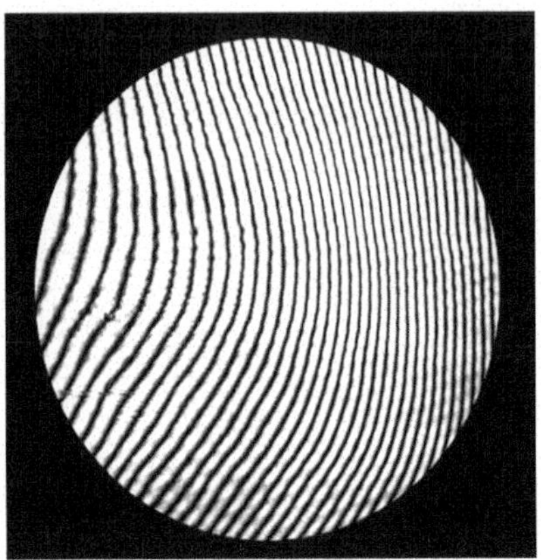

Fig. 8.4 Interferogram registered for Fourier reconstruction

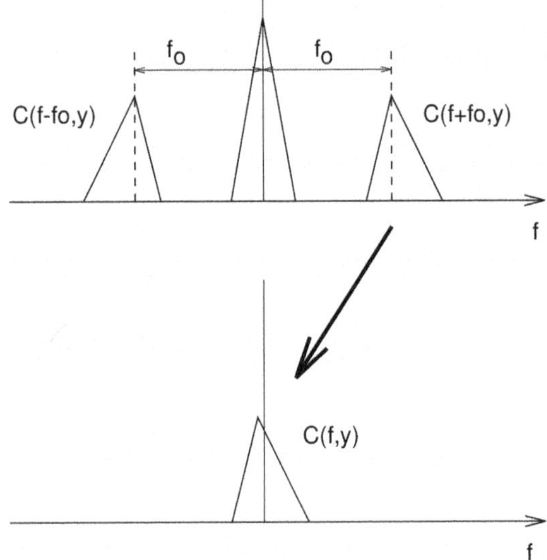

Fig. 8.5 Fourier spectra of a fringe pattern modulated by spatial frequency f_o (top) and a single spectrum translated to the origin (bottom)

Then one of the side-lobe spectra $C(f + f_o, y)$ or $C^*(f - f_o, y)$ is shifted to the origin, as shown in Fig. 8.5. The inverse Fourier transform then yields $c(x, y)$ as defined in Eqs. (8.12, 8.13). Taking the complex logarithm gives the phase $\psi(x, y)$:

$$i\psi(x, y) = \ln[c(x, y)]. \tag{8.15}$$

The phase $\psi(x, y)$ is obtained modulo 2π, with principal values in the range $-\pi \ldots \pi$. To recover a continuous phase map, phase unwrapping is applied. In its simplest form, unwrapping removes $\pm 2\pi$ discontinuities at phase jumps. For noisy two-dimensional data, general phase unwrapping is an ill-posed problem.

The Fourier method has limitations:

- The minimum tilt of the reference wavefront, relative to the sample wavefront, must exceed the maximum tilt in the sample wavefront. This condition is equivalent to the absence of closed fringes in the recorded pattern.
- The maximum tilt of the reference wavefront in the registration plane is limited by the pixel pitch P of the image sensor. According to the sampling theorem, more than two pixels must correspond to one fringe period at maximum fringe frequency. Thus, the maximum allowable tilt φ is:

$$\varphi < \frac{\lambda}{2P}. \tag{8.16}$$

If the fringe pattern is imaged to the sensor by an optical system with magnification $M < 1$, the maximum tilt in the object plane is

$$\varphi < \frac{M\lambda}{2P}. \tag{8.17}$$

Fourier reconstruction is the method of choice in laboratory experimental practice, as it implies a very simple hardware realization.

8.6 Shearing Interferometers

In a shearing interferometer, a wave interferes with a shifted and delayed copy of itself. If the lateral shift is δ and the wavefront tilt is α, then the fringe corresponds to a tilt of

$$\alpha = \frac{m\lambda}{\delta}, \tag{8.18}$$

where m is the fringe order. A tilted flat wavefront produces uniform color across the field.

Unlike in classical interferometers, where fringe frequency is proportional to tilt, in a shearing interferometer fringe frequency is proportional to wavefront curvature. For a curved wavefront with radius of curvature R, the fringe positions are found from

$$\frac{a^2}{2R} - \frac{(a+\delta)^2}{2R} = m\lambda, \tag{8.19}$$

with the solution

$$a_m = \frac{2Rm\lambda + \delta^2}{2\delta}, \tag{8.20}$$

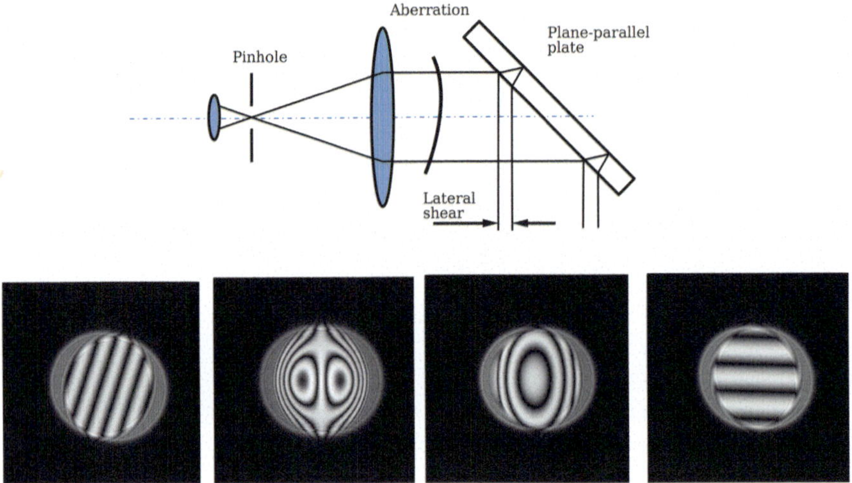

Fig. 8.6 Shearing interferometer and fringe patterns for defocus, spherical aberration, coma, and astigmatism (left to right)

and spatial fringe period

$$T = \frac{R\lambda}{\delta}. \tag{8.21}$$

The method requires a source with spatial coherence length exceeding δ, and the optical path difference introduced by the shearing plate must be smaller than the coherence length. The lateral resolution of the interferometer is limited to δ. From Eq. 8.19 it follows that increasing angular resolution reduces spatial resolution, leading to a relation analogous to the uncertainty principle: $\delta\alpha = \lambda$.

The configuration in Fig. 8.6 is only one example; many interferometer configurations exist to produce self-interfering beams. A simple method, proposed by Ronchi [4], uses a rough grating near the focus of a converging beam to generate fringes by interference of diffracted beam copies.

A single fringe pattern obtained with a shearing interferometer is sensitive only to aberrations along the shear direction. For full wavefront reconstruction, two fringe patterns with orthogonal shears (along x and y axes) must be recorded (Fig. 8.6).

8.7 Hartmann–Shack Sensors

The Hartmann–Shack (HS) sensor consists of an array of subapertures placed in the image of the entrance pupil of the optical system as shown in Fig. 8.7. Each subaperture contains a microlens to forms a light spot on an image sensor positioned

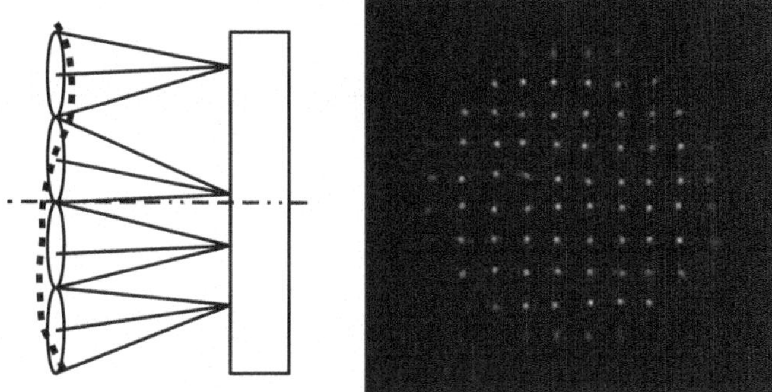

Fig. 8.7 Schematic (left) and spot pattern obtained with a HS sensor (right)

behind the array. A typical spot pattern produced with the HS sensor is shown in Fig. 8.7.

If the wavefront is flat, the centroids of the light spots coincide with the regular subaperture positions. When the wavefront is distorted, the spot positions deviate due to local wavefront tilts within the subapertures. Although each light spot may be distorted by higher-order aberrations within its subaperture, these distortions are usually neglected. Only the lateral shifts of the centroids, caused by local wavefront tilts, are taken into account. In this framework, the displacement of a centroid $\bar{\Delta}(x, y)$ is proportional to the local wavefront tilt $\nabla\varphi(x, y)$ and to the distance L between the lenslet array and the image sensor.

8.7.1 Modal Wavefront Reconstruction

An HS sensor with N subapertures produces a vector of $2N$ spot shifts,

$$\bar{s} = \begin{vmatrix} \bar{s_x} \\ \bar{s_y} \end{vmatrix},$$

where N values correspond to $\bar{s_x}$ and N to $\bar{s_y}$. The wavefront is reconstructed as a linear combination of M basis functions $\psi_i, \quad i = 1 \dots M$:

$$\varphi = \sum_{i=1}^{M} a_i \psi_i. \tag{8.22}$$

The response vector $\bar{s}$ is then defined as:

$$s_{xj} = \sum_{i=1}^{M} a_i \frac{\partial \psi_i(x_j, y_j)}{\partial x},$$
$$s_{yj} = \sum_{i=1}^{M} a_i \frac{\partial \psi_i(x_j, y_j)}{\partial y}.$$

The sensor is calibrated by measuring or calculating the matrix $\bar{r}$, whose columns are the sensor responses to each basis function applied with unit coefficient:

$$\bar{r}_i = \begin{vmatrix} \frac{\partial \psi_i(x_j, y_j)}{\partial x} \\ \frac{\partial \psi_i(x_j, y_j)}{\partial y} \end{vmatrix}. \quad (8.23)$$

The reconstruction problem is reduced to solving the linear system

$$|r|\bar{a} = \bar{s}, \quad (8.24)$$

where $\bar{s}$ is the measured response to an unknown aberration, $|r|$ is the calibration matrix, and $\bar{a}$ is the vector of expansion coefficients in Eq. 8.22.

In most cases $M \neq 2N$. Typically, the number of subapertures N in the HS sensor is larger than the number of basis functions: $N > M$. Since an exact solution of Eq. 8.24 is then impossible, we minimize the error

$$\epsilon = (|r|\bar{a} - \bar{s})(|r|\bar{a} - \bar{s}),$$
$$\epsilon = \bar{a}^T |r|^T |r|\bar{a} - 2\bar{s}^T |r|\bar{a} + \bar{s}^T \bar{s}.$$

To find the extremum of ϵ with respect to $\bar{a}$, we require $d\epsilon/da = 0$:

$$2|r|^T |r|\bar{a} - 2|r|^T \bar{s} = 0.$$

Finally, the least-squares solution is given by

$$\bar{a} = (|r|^T |r|)^{-1} |r|^T \bar{s}. \quad (8.25)$$

The operator $(|r|^T |r|)^{-1} |r|^T$ defines the pseudoinverse of the matrix $|r|$.

In the absence of noise, the solution 8.25 provides the best-fit decomposition of the sensor response as a sum of derivatives of the basis functions.

Two special cases of local point-wise phase reconstruction at the lenslet coordinates in a rectangular HS array are described in [5] and [6].

Reconstruction from HS patterns is reduced to matrix multiplication, and can be performed in real time. This speed advantage is widely used in the control systems for real-time adaptive optics.

8.8 Intensity Transport

When a beam is focused and converges, its intensity increases during propagation. Conversely, when the beam diverges due to defocus, its intensity decreases as shown in Fig. 8.8. The relation between wavefront curvature and intensity modulation in a propagating coherent beam is described by the intensity transport equation.

In the parabolic approximation, wave propagation is governed by Eq. 14.6:

$$\frac{\partial^2 U}{\partial x^2} + \frac{\partial^2 U}{\partial y^2} + 2ik\frac{\partial U}{\partial z} = 0. \tag{8.26}$$

To derive the intensity transport equation, we express the amplitude U in Eq. 8.26 through the beam intensity $I(z)$ and phase $\varphi(x, y)$:

$$U = \sqrt{I(z)}\exp(i\varphi(x, y)). \tag{8.27}$$

Substituting Eq. 8.27 into Eq. 8.26 and separating real and imaginary parts yields the transport of intensity equation:

$$\frac{\partial^2 \varphi}{\partial x^2} + \frac{\partial^2 \varphi}{\partial y^2} = -\frac{k}{I}\frac{\partial I}{\partial z}. \tag{8.28}$$

The Poisson Eq. 8.28 can be solved for φ once the intensity distributions in two planes, at z and $z + \Delta$, are registered.

However, the solution of Eq. 8.28 excludes all functions with $\nabla^2\varphi = 0$ and depends on the boundary conditions. Consider three distributions: $\varphi_1 = x^2, \varphi_2 = y^2$, and $\varphi_3 = (x^2 + y^2)/2$. In all three cases $\nabla^2\varphi_1 = \nabla^2\varphi_2 = \nabla^2\varphi_3 = 2$. This degeneracy can only be resolved by imposing appropriate boundary conditions on the solution. The method is therefore useful for controlling higher-order aberrations, but

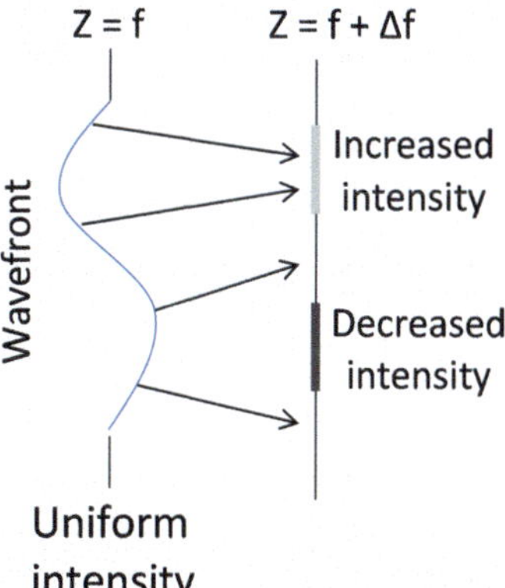

Fig. 8.8 Illustration of the intensity transport equation

must be combined with another technique, such as HS sensing or interferometry, to define the boundary conditions for low-order aberrations.

Accurate reconstruction requires linear registration of intensity. Measurement noise can produce nonlinear effects in low-intensity regions, even when using a linear detector.

Finally, spatial coherence of light is crucial, since Eq. 8.28 is derived under the assumption of spatial coherence. For reliable results, the coherence area should be larger than the beam aperture.

8.9 Conclusion

Wavefront aberrations, surface roughness, and localized defects affect the quality of optical systems and need to be strictly controlled. The Strehl parameter provides a compact criterion for diffraction-limited performance by linking wavefront error to focal intensity.

Surface roughness directly influences scattering. Its characterization relies on standardized parameters such as R_q and R_z, and the total integrated scatter. Mirrors must satisfy stricter surface requirements than lenses. For short-wavelength systems, grazing incidence reduces scattering and enables high reflectivity, as in X-ray telescopes.

Defects such as scratches and digs are classified by international standards, with relaxed values sufficient for large-aperture imaging. High power laser components demand extremely low defect densities to avoid damage.

Testing methods vary according to the scale and nature of the required information. Stylus profilometry measures surface figures and roughness with high spatial resolution, suitable for micro-optics and diffractive elements. Interferometry remains the most used technique for optics control. Self-referencing shearing interferometers provide diagnostics of curvature and aberrations without external references, though with reduced spatial resolution and rather high demands to beam coherence. Hartmann–Shack sensors sample the wavefront slopes and enable local and modal reconstruction of the wavefront, with real-time applications in adaptive optics. Intensity transport methods link the intensity modulations to wavefront curvature, though some aberrations can not be sensed with curvature methods.

Together, these methods provide a set of tools for quantifying optical quality, ensuring the specifications for imaging, laser, and precision instruments are met.

8.10 Problems

8.1 For an ideal lens with focal length F and focal number $F\#$ define maximum value of defocus ΔF at which the Strehl number S satisfies $S > 1/e$ at wavelength λ. Compare this value with the diffraction-limited depth of focus.

Solution (8.1). *In the parabolic approximation, the aberration $a(\rho)$ due to defocus ΔF is given by*

$$a \approx -\frac{\rho^2 \Delta F}{2\,F^2},$$

and the maximum aperture equals to $F\#/F/2$. The rms aberration can be calculated as

$$\sigma^2 = \frac{1}{(F\#/F/2)^2} \int_0^{F\#/F/2} \left[\frac{\rho^2 \Delta F}{2\,F^2}\right]^2 \rho d\rho = \frac{\Delta F^2}{384\,F\#^4}.$$

solving the equation

$$\exp(-k\sigma^2) > \exp(-1),$$

for ΔF with $k = 2\pi/\lambda$, we obtain

$$|\Delta F| < 3.2\,F\#^2\lambda,$$

which is pretty close to the value of diffraction-limited depth of focus defined by the expression 2.28.

8.2 What is the necessary minimum number of subapertures in a Hartmann-Shack sensor for detection of defocus aberration?

Solution (8.2). *3 subapertures, positioned in a triangular pattern.*

8.3 For a shearing interferometer with thickness d refraction index n and incidence angle α, derive the transversal shift Δ. What is the optical path difference δL between the two beams?

Solution (8.3). *In the paraxial approximation, the shift equals*

$$\Delta = \frac{2d\alpha}{n},$$

and the delay, expressed as optical path difference is

$$\delta L \approx n\sqrt{4d^2 + \Delta^2}$$

8.4 Formulate the requirements for the coherence of light to obtain clear interference pattern in a shearing interferometer formed by a glass plate with the thickness d, refraction index n, at the incidence angle of α.

Solution (8.4). *The spatial coherence size should be larger than* Δ*, and the coherence length should be larger than the path difference* δL*, both defined in the solution to problem 8.3.*

8.5 Estimate the expected focus error, in diopters, obtained with control of a collimated beam with $\lambda = 0.5\ \mu$m, beam diameter D of 1 cm, using shearing interferometer providing lateral shift of $\delta = 1$ mm.

Solution (8.5). *In the assumption that for uniform color over the whole aperture, the phase difference should not exceed* $\lambda/8$*, we will use the estimate* $D = T/8$ *where* T *is the fringe period as in 8.21. Then, using 8.21 we have*

$$8D = \frac{R\lambda}{\delta},$$

and

$$R = \frac{8D\delta}{\lambda} = 160\ m.$$

The estimate for the focus error is $\sim 1/160$ *diopter.*

8.6 Draw the X and Y cross sections of the wavefronts for interferograms shown in Fig. 8.1. Do the same after rotating the X-Y coordinate system by 45°.

8.7 Describe the function of the positive lens in the Fizeau interferometer shown in Fig. 8.2. Calculate the focal length of this lens if the distance from the imaging sensor to the lens is a, and the distance from the lens to the test flat, measured along the optical axis, is b.

References

1. H.E. Bennett, J.O. Porteus, Relation between surface roughness and specular reflectance at normal incidence. J. Opt. Soc. Am. **51**(2), 123–129 (1961). https://doi.org/10.1364/JOSA.51.000123
2. C.Y. Poon, B. Bhushan, Comparison of surface roughness measurements by stylus profiler, AFM and non-contact optical profiler. Wear **190**(1), 76–88 (1995)
3. M. Takeda, H. Ina, S. Kobayashi, Fourier-transform method of fringe-pattern analysis for computer-based topography and interferometry. J. Opt. Soc. Am. **72**(1), 156–160 (1982). https://doi.org/10.1364/JOSA.72.000156
4. V. Ronchi, Due nuovi metodi per lo studio delle superficie e dei sistemi ottici. Il Nuovo Cimento (1911–1923) **26**(1), 69–71 (1923). https://doi.org/10.1007/BF02959347
5. R.H. Hudgin, Wave-front reconstruction for compensated imaging. J. Opt. Soc. Am. **67**(3), 375–378 (1977). https://doi.org/10.1364/JOSA.67.000375
6. D.L. Fried, Least-square fitting a wave-front distortion estimate to an array of phase-difference measurements. J. Opt. Soc. Am. **67**(3), 370–375 (1977). https://doi.org/10.1364/JOSA.67.000370

Chapter 9
Radiometry, Light Sources and Sensors

Abstract Photometry and radiometry study the energy transport of visible light (photometry) and the general laws of energy transport in optical systems (radiometry). Brief introduction is given to radiometric and photometric units. Propagation of energy from source to image as a functional of the parameters of an imaging system is described and expressions governing the illuminance of optical image are derived. A brief introduction to the sources and sensors of light is given, the major parameters of light sources and sensors are defined providing basic expressions for radiometric calculations in optical systems with different kinds of sources and sensors.

9.1 Radiometry and Photometry

Both the *radiometric* and *photometric* systems are used to describe energy transport in optics.

In the radiometric system, traditional physical quantities such as Joule, Watt, meter, and steradian are applied to light measurements. The radiometric system is not related to human perception of light.

In the photometric system, light is measured in special units: lumen, lux, and candela. These units are weighted by the sensitivity of the human eye. The photometric system is therefore used for measurements related to visual perception. A good introduction to radiometry and photometry can be found in [1, 2].

9.1.1 Radiometric Quantities

The energy of light is measured in Joule [J]. The power of light, also called *flux* Φ, is measured in Watt [W]. The density of optical power E incident on a surface, or emitted by a surface, is measured in [W/m^2]. This quantity is also called *flux density* or *irradiance*. For emitting surfaces the term *radiosity* is often used.

G. Vdovin, *Elementary Technical Optics*, UNITEXT for Physics,
https://doi.org/10.1007/978-3-032-08626-6_9

Since a light beam is always limited by a certain solid angle, the angular density of optical power, called *radiant intensity I* (or simply *intensity*), is measured in Watt per steradian [W/sr].

Radiance (also referred to as *brightness*—a term deprecated in scientific literature but still common in engineering practice) L characterizes a source emitting a given power [W] from an elemental area [m^2] into an elemental solid angle [sr]. Radiance is expressed in [W/($m^2\cdot$ sr)].

9.1.2 Photometric Quantities

Photometric power is measured in *lumen*. The conversion between radiometric and photometric units is defined by Eq. 9.2 and depends on the spectral composition of light. Analogous to irradiance, photometric illuminance is measured in lumens per square meter, with the unit *lux*. The light intensity, measured in lumens per steradian, has the unit *candela*. Brightness, or luminance, is measured in candela per square meter, with the unit *nit*. The unit candela per square centimeter, although rarely used today, is called *stilb*.

The conversion between the two systems is obtained by integrating the radiometric quantity weighted by the spectral sensitivity of the human eye. The International Commission on Illumination (CIE) defines the standard spectral sensitivity function $V(\lambda)$, published in 1978. The function has its maximum at 555 nm and corresponds to color (photopic) vision. For low illumination levels (scotopic vision), the maximum is at 507 nm. The lumen is defined as light power weighted by $V(\lambda)$. The maximum conversion coefficient between lumens and Watts is:

$$1\ \text{W} = 683\ \text{lm} \quad \text{at}\ \lambda = 555\ \text{nm}.$$

For engineering estimates, $V(\lambda)$ can be approximated by a Gaussian curve:

$$V(\lambda) = 1.019 \exp(-285.4(\lambda - 0.559)^2). \tag{9.1}$$

The difference between this approximation and the CIE standard is shown in Fig. 9.1. For accurate work, the tabulated CIE curve must be used.

For a multi-color light beam with spectral density $D(\lambda)$ [W/μm], the photometric power in lumens is calculated as

$$\Phi = 683 \int_{0.38}^{0.83} V(\lambda) D(\lambda)\, d\lambda. \tag{9.2}$$

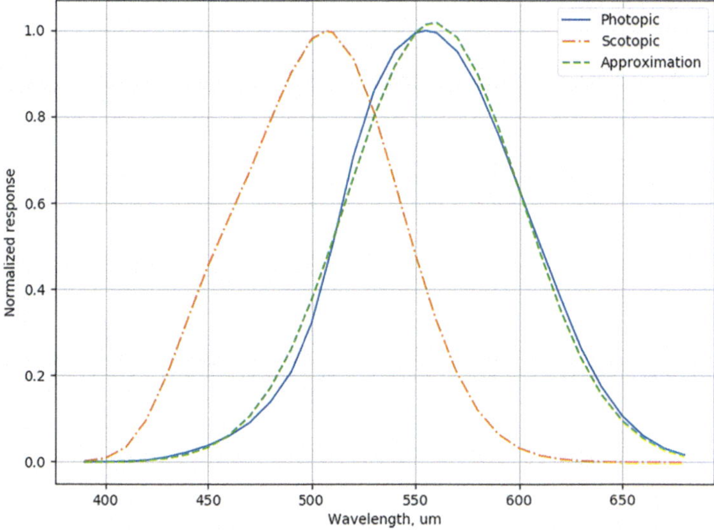

Fig. 9.1 Normalized eye sensitivity curves. The "Approximation" curve is calculated using Eq. 9.1

9.1.3 Definition of Solid Angle

Assume a light beam with arbitrary cross-section is emitted from a source placed at the center of a unit sphere. The solid angle subtended by the beam equals the area of the illuminated patch on the sphere. A 4π solid angle corresponds to a full sphere, and a 2π solid angle corresponds to a full hemisphere.

For a cone with apex angle 2θ, the solid angle is

$$A = 4\pi \sin^2\left(\frac{\theta}{2}\right) = 2\pi(1 - \cos\theta). \tag{9.3}$$

For the practically important case $\theta \leq 1$, the approximation

$$A \approx \pi\theta^2 \tag{9.4}$$

is widely used.

9.2 Etendue and Radiation Transfer

As introduced in Sect. 3.2, the etendue is defined as $\epsilon = n^2\, dS\, dA$, where dS is the area and dA is the solid angle, both measured in the same plane. The etendue remains invariant throughout the optical system.

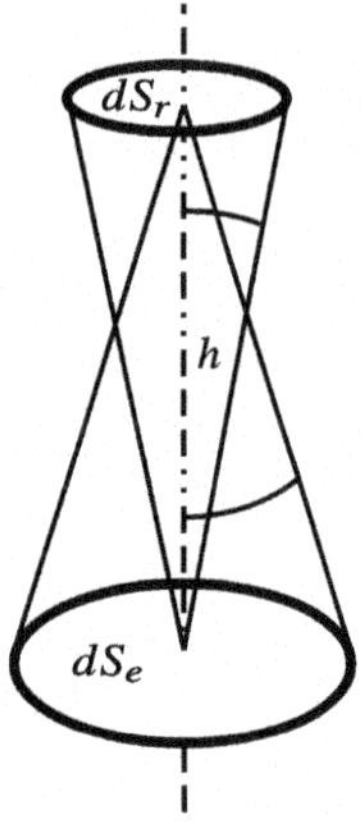

Fig. 9.2 Illustration of radiation transfer from an emitter with area dS_e to a receiver with area dS_r at distance h

Consider a plane emitter of brightness L with an elementary area dS_e, orthogonal to the optical axis. It illuminates a receiver with elementary area dS_r, also orthogonal to the optical axis, positioned at a distance h from the emitter (Fig. 9.2).

The flux $d\Phi$ through the receiver dS_r is given by the product of source brightness L, the solid angle $\frac{dS_e}{h^2}$ subtended by the emitter area dS_e, and the receiver area dS_r:

$$d\Phi = L\,dS_r\frac{dS_e}{h^2} = L\,dS_e\frac{dS_r}{h^2}. \tag{9.5}$$

Since the terms $dS_r\frac{dS_e}{h^2}$ and $dS_e\frac{dS_r}{h^2}$ have the same physical meaning as the etendue invariant (neglecting refraction coefficients for simplicity), the total flux through the system can be expressed as the product of source brightness and etendue.

9.2.1 Irradiance of an Optical Image

To calculate the irradiance of an image of an object with brightness L formed by an optical system, consider Fig. 9.3. A disk-shaped object of brightness L and angular size $d\varphi$ is imaged by a lens of diameter D. The lens forms the image at distance F from its back principal plane.

The solid angle subtended by the object is $\pi(d\varphi)^2$. According to 9.5, the total flux Φ received by the lens is

$$\Phi = L\frac{\pi D^2}{4}\pi(d\varphi)^2.$$

This flux is uniformly distributed over the image area $S = \pi(d\varphi F)^2$. Therefore, the image irradiance is

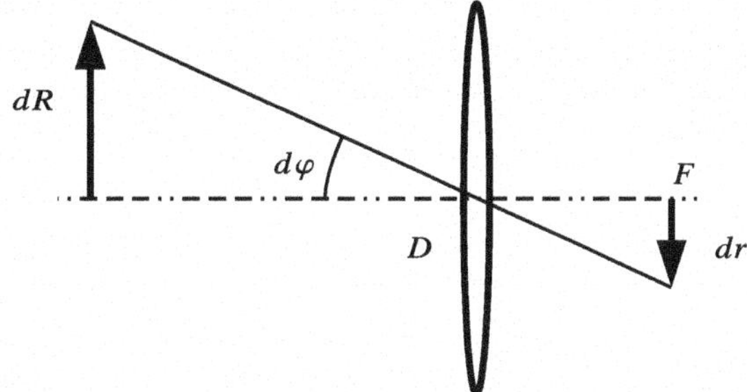

Fig. 9.3 Derivation of the irradiance of an optical image

$$E = \frac{\Phi}{S} = L\frac{\pi D^2}{4S'^2}. \tag{9.6}$$

In this derivation S' denotes the image distance. In many practical cases, when the object distance is large compared to the focal length, S' can be approximated by the focal length F. Then Eq. 9.6 becomes

$$E = L\pi A^2 \approx L\frac{\pi}{4(F\#)^2}, \tag{9.7}$$

where $F\# = F/D$ is the f-number, and the numerical aperture is $A = (F\#^2 + 1)^{-1/2}$. The image irradiance is thus proportional to the object brightness and to the square of the numerical aperture in the image space. This aperture may differ from the nominal NA of the lens, especially in macro mode when the image distance S' is much larger than the focal length. Losses due to absorption in the lens material and propagation should also be considered.

For a circular Lambertian patch of brightness L directly illuminating the receiver, the irradiance can be written similarly to 9.7:

$$E = L\pi \sin^2(\varphi), \tag{9.8}$$

where φ is the half-apex angle subtended by the surface. For an infinite surface, $\varphi = \pi/2$, and

$$E = L\pi. \tag{9.9}$$

9.2.2 *Cosine Law*

Most emitting and reflecting sources can be approximated as Lambertian. For a Lambertian source, the brightness L [W/(m^{2}sr)] is constant in all directions. However, this does not imply that the radiant intensity I [W/sr] is constant. As shown in Fig. 9.4,

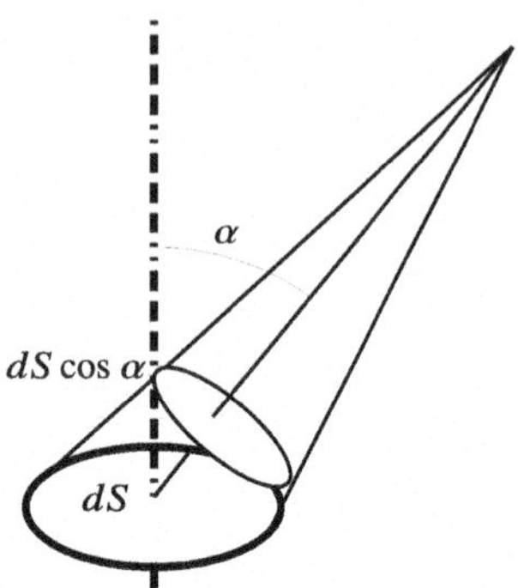

Fig. 9.4 Illustration to Lambert's cosine law

the radiant intensity I is proportional to the cosine of the angle α between the surface normal and the observation direction:

$$I = L dS \cos(\alpha). \tag{9.10}$$

It follows that the illuminance E of a flat surface illuminated by a beam of incident irradiance E_i [W/m^2] at angle α is

$$E = E_i \cos(\alpha). \tag{9.11}$$

9.2.3 *Light Falloff in a Lens*

Lambert's cosine law 9.10 explains how image irradiance changes with the angular position α of the object:

- The object intensity in the lens direction is proportional to $\cos(\alpha)$.
- The irradiance at the lens pupil is proportional to $\cos(\alpha)$.
- The intensity at the exit pupil is proportional to $\cos(\alpha)$.
- The irradiance in the image plane is proportional to $\cos(\alpha)$.

Taken together, these effects imply that, in the absence of vignetting and aberrations, the image irradiance decreases as the fourth power of the cosine of the field angle α relative to the optical axis:

$$E \sim E_0 \cos^4(\alpha). \tag{9.12}$$

Most conventional lenses follow this rule. Wide-angle lenses with field of view approaching or exceeding $\pi/2$ require special designs. According to Eqs. 9.11 and 9.12, the image irradiance at $\alpha = \pi/2$ should vanish. In practice, this falloff is mitigated by specific design features, such as hemispherical front elements and strong distortion. An example is shown in Fig. 9.5.

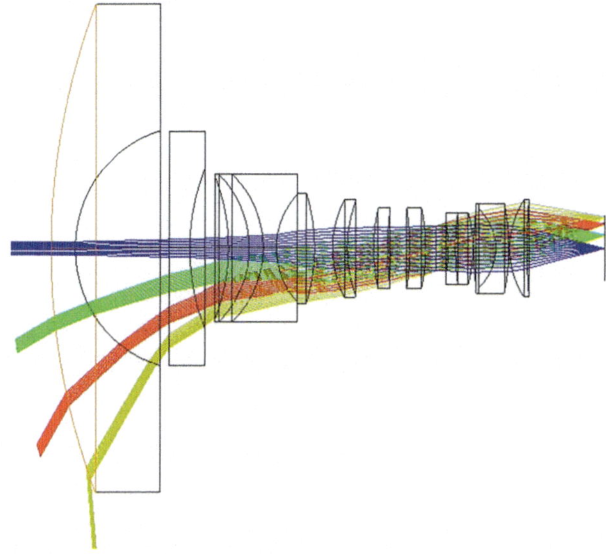

Fig. 9.5 Wide-angle lens with field of view of 105°

9.2.4 Blackbody Emission

Any body at nonzero temperature emits electromagnetic radiation. The total irradiance E [W/m^2] emitted by a blackbody is described by the Stefan–Boltzmann law:

$$E = \sigma T^4, \tag{9.13}$$

where $\sigma = 5.67 \cdot 10^{-8}$ [W m^{-2} K^{-4}] is the Stefan–Boltzmann constant. A blackbody is defined as an emitter with unity emissivity at all wavelengths. Real emitters have lower emissivity and are described as gray bodies.

The spectral irradiance E_λ [W/(m^2·m)] of a blackbody at temperature T is given by the Planck's law:

$$E_\lambda = \frac{2\pi hc^2}{\lambda^5 \left(e^{hc/(\lambda k_b T)} - 1\right)}, \tag{9.14}$$

where $h \approx 6.626 \cdot 10^{-34}$ J· s is Planck's constant, $k_b \approx 1.38 \cdot 10^{-23}$ J/K is the Boltzmann constant, and c is the speed of light.

The maximum spectral emissivity occurs at wavelength λ_{max}, defined by Wien's displacement law:

$$\lambda_{max} = \frac{b}{T}, \tag{9.15}$$

where $b \approx 2898$ [μm· K] is Wien's displacement constant.

For a flat black surface of area S, emitting with brightness L into the solid angle 2π, the total emitted power I is

$$I = SL\pi, \tag{9.16}$$

where the factor π (not 2π) accounts for the cosine-law correction over a hemisphere.

In thermal equilibrium, either within a closed system or between such systems, emission (Eqs. 9.13 and 9.16) equals absorption. Any additional incoming radiation increases temperature until a new equilibrium is reached. It implies that no optical system imaging a thermal source can produce a focal temperature higher than that of the source itself.

9.2.5 *Light Emitting Diodes (LEDs)*

A Light Emitting Diode (LED) is a semiconductor device that produces light through quantum electroluminescence. When current passes through a semiconductor diode, electrons and holes recombine in the p–n junction region. The energy released at each recombination is defined by the bandgap width. This energy may dissipate as heat or be emitted as a photon, depending on the electronic band structure.

Radiative recombination requires carrier momentum near zero. If the momentum is not close to zero, recombination is non-radiative and no photon is produced. This momentum conservation rule excludes Si and Ge as efficient light emitters, limiting practical materials to GaAs, InGaN, and GaN. The emission wavelength is determined by the bandgap E_g as

$$\lambda \approx \frac{1.242}{E_g}, \tag{9.17}$$

where λ is in microns and E_g is in electron-volts.

The LED emission spectrum is relatively broad (tens of nanometers for visible LEDs) due to the extended structures of energy levels. Photons are emitted in many directions, limiting directionality. Because semiconductors have high refractive indices, many photons undergo total internal reflection at the semiconductor–air interface. To increase the TIR angle and improve extraction efficiency, LEDs are encapsulated in plastic domes of higher refractive index.

LEDs are available across a broad range, from IR to UV. White LEDs are typically made by coating a blue or UV LED with a broadband phosphor, which re-emits across the visible spectrum. Alternatively, white light can be produced by combining red, green, and blue LEDs in a single package.

9.3 Light Sensors

Light sensors convert optical energy into information signals, usually—though not exclusively—electric. A number of physical effects can be exploited for light sensing. A good overview of the basic principles is given in [3].

9.3.1 *Thermal Sensors*

Thermal sensors convert temperature changes into electrical signals.

Bolometers rely on the dependence of the conductivity of a metal or semiconductor on temperature. Arrays of bolometers are the most widely used uncooled image sensors for the far-IR range $\lambda = 3 \ldots 12\ \mu\text{m}$.

The modulation ΔR of bolometer resistance depends on the change in temperature ΔT, which in turn is modulated by the absorbed flux Φ: $\Delta T = \Phi\tau/C$, where C is the thermal capacitance of the bolometer and τ is the flux duration. The resulting resistance change is

$$\frac{\Delta R}{R} = \beta \Delta T, \tag{9.18}$$

where β is the thermal coefficient of resistance. For most metals $\beta \approx \frac{1}{T}$, while for semiconductors $\beta \approx -\frac{K}{T^2}$, with $K = 3000 \ldots 10000$.

Thermocouples and Peltier sensors use the thermoelectric effect. They are primarily used for temperature sensing but can also detect light through its heating effect. A Peltier sensor with active cooling can measure light beams of very high power, up to several kW of continuous-wave radiation.

Pyroelectric sensors generate voltage when subjected to a non-stationary thermal field, such as that caused by pulsed radiation.

Calorimeters have very low temporal resolution but allow measurement of large beam energies by absorbing the light in a thermally isolated body with calibrated thermal capacity and measuring the resulting temperature rise.

9.3.2 *Photoelectric Sensors*

The photoelectric effect is the generation of free carriers by direct action of light.

In *vacuum photoelements, photomultipliers, and image intensifiers*, free electrons are generated by the interaction of photons with a photocathode. These electrons are accelerated in vacuum toward an anode by an electric field, either directly or via intermediate electrodes (dynodes) that provide secondary emission, multiplying the number of electrons. Photomultipliers are used in applications requiring single-photon counting. Image intensifiers use magnetic or electrostatic lenses to focus

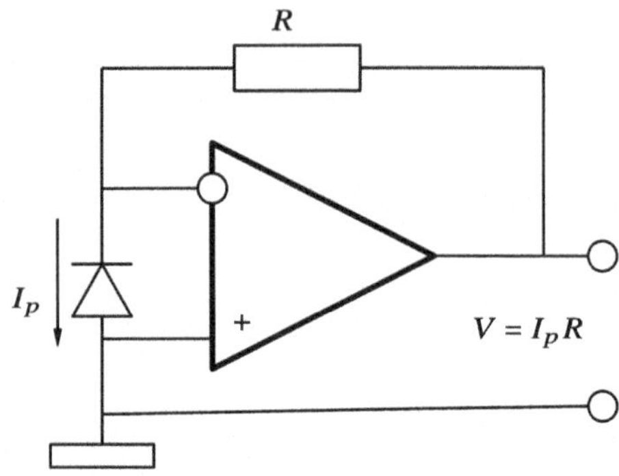

Fig. 9.6 Typical photodiode bootstrap with an op-amp

emitted photoelectrons onto a phosphor screen, producing an amplified optical image of faint objects.

In *photoresistors*, absorbed light generates additional carriers, reducing the sensor impedance.

In *photodiodes*, electron–hole pairs are generated in a p–n junction, producing a current proportional to the flux. The spectral sensitivity is limited by the bandgap E_b, with the maximum detectable wavelength given by (cf. 9.17):

$$\lambda_{max} = \frac{1.242}{E_b}, \tag{9.19}$$

where E_b is measured in eV. The open-circuit voltage of a photodiode increases exponentially with incident flux until saturating at the forward bias. By contrast, the short-circuit current is directly proportional to the flux over a wide range. In this mode, the responsivity of a typical photodiode lies in the range $0.1 \ldots 0.6$ [A/W].

A typical signal-conditioning circuit for a photodiode, ensuring a linear relation between optical flux and output voltage, is shown in Fig. 9.6. In this configuration, the photodiode operates at zero bias, providing a linear voltage response $V = I_p R$. Photodiodes integrated with readout electronics are the basis of *CMOS image sensors*. More details on photodiode signal conditioning can be found in [4]. In *charge-coupled devices (CCDs)*, photogenerated carriers are stored in a periodic MOS structure, and then cyclically transferred from pixel to pixel toward the electronic readout.

9.3.3 Noise in Photodetectors

Shot noise arises from the discrete nature of photons. Assuming Poisson statistics, for uncorrelated photon arrivals the probability of detecting n photons in a time interval τ is

$$p(n) = \frac{N^n e^{-N}}{n!}, \tag{9.20}$$

where N is the average number of photons registered during τ. The photon noise corresponds to the *rms* deviation of photon number from the mean:

$$\sigma_N = \sqrt{\sum_0^\infty (N - n)^2 p(n)} = \sqrt{N}. \tag{9.21}$$

The signal-to-noise ratio (SNR) is therefore

$$\text{SNR} = \frac{N}{\sigma_N} = \sqrt{N}. \tag{9.22}$$

Photon noise dominates at very low light levels or very short measurement times, when N is small and SNR is correspondingly low.

Thermal noise arises from thermal motion of electrons and can be modeled as a noise current source in parallel with the sensor:

$$i_n = \sqrt{\frac{4k_B T \Delta f}{R}}, \tag{9.23}$$

where $k_B \approx 1.381 \cdot 10^{-23}$ [J/K] is Boltzmann's constant, T is absolute temperature, Δf is the bandwidth, and R is sensor resistance. When short-circuited, the sensor dissipates noise power

$$P_n = i_n^2 R = 4k_B T \Delta f. \tag{9.24}$$

Since $P_n = v_n^2/R$, the equivalent noise voltage is

$$v_n = \sqrt{4k_B T R \Delta f}. \tag{9.25}$$

The noise voltage increases with source resistance; sources with high output impedance generate more voltage noise.

9.3.4 Noise Equivalent Power and Detectivity

Sensor sensitivity is characterized by responsivity $\Re$ and noise equivalent power (NEP).

The responsivity $\Re$ [A/W] is the current produced per unit of absorbed optical power:

$$\Re = \frac{i}{P}. \tag{9.26}$$

Substituting $i = \eta e \Phi$ and $P = h\nu\Phi$, where η is quantum efficiency, Φ is the photon flux [photons/s], e is the electron charge, h is Planck's constant, and ν is the photon frequency, gives

$$\Re = \frac{\eta e \Phi}{h\nu\Phi} = \frac{\eta e \lambda}{hc} = \eta \frac{\lambda[\mu\mathrm{m}]}{1.24} \left[\frac{\mathrm{A}}{\mathrm{W}}\right]. \tag{9.27}$$

NEP is the signal power required to achieve SNR = 1 in a 1 Hz bandwidth. It is measured in [W/Hz$^{1/2}$].

The detectivity $D = 1/\mathrm{NEP}$ [Hz$^{1/2}$/W] is the reciprocal of NEP. It represents the SNR achieved with 1 W input power in a 1 Hz bandwidth.

Since detectivity depends on sensor area, a normalized detectivity D^* is used:

$$D^* = \frac{\sqrt{A\Delta f}}{\mathrm{NEP}}, \tag{9.28}$$

where A is the sensor area and Δf is the bandwidth. D^* is measured in [m Hz$^{1/2}$/W].

Detectivity depends both on quantum efficiency η–the fraction of incident photons that generate signal–and on internal noise sources in the sensor.

9.4 Conclusion

Radiometric quantities describe optical energy transport independent of human perception. Photometric quantities are weighted by the spectral sensitivity of the human eye and are obtained by integrating radiometric distributions with the function of spectral sensitivity of the human eye.

Image irradiance in the optical system is determined by the object brightness, angular size, and the numerical aperture of optic. The law of $\cos^4$ defines the field-dependent irradiance falloff in conventional lenses. Lambert's law describes the angular distribution of radiant intensity from diffuse emitters.

Blackbody emission defines the spectrum of a thermal source. Semiconductor light sources operate through radiative recombination in p–n junctions, with emission wavelength defined by the bandgap. LEDs provide high efficiency and broad spectral coverage.

Thermal sensors, including bolometers, pyroelectric sensors, and calorimeters, are suited for infrared and high-energy flux measurement. In photoelectric sensors photons directly activate charge carriers. Photo-elements and photomultipliers are used for high sensitivity in photon-counting applications. Potodiodes, CCD and CMOS arrays are used for imaging.

Noise performance of a radiation sensor is defined by a number of effects, including photon shot noise and thermal current fluctuations. Noise equivalent power and detectivity describe the sensor sensitivity across bandwidth and area, as a function of quantum efficiency and internal noise.

9.5 Problems

9.1 The illuminance of the sensor, created by the star image in a telescope is 10 times weaker than the illuminance created by the background sky. How the diameter of telescope D should be changed to make the illuminance produced by star 10 times stronger than the background?

Solution (9.1). *The sky irradiance is proportional to D^2, the point source image irradiance in the focus is proportional to D^4. The diameter D should be made 10 times larger.*

9.2 Photographer uses camera lens with $F = 5$ cm and $F\# = 2$ to make a photo of a remote object from a distance of 100 m. Then he decides to make a closeup, approaches the object and makes a sharp macro photograph at 1:1 scale using the same lens with the same $F\#$ setting. How the illuminance of the image will differ between the first and the second photograph?

Solution (9.2). *In the first case the object is at infinity, the image is in focus of the lens. When the object is imaged with a scale of 1:1, the image is obtained at double focal length from the principal plane, therefore the F# is doubled. The illuminance of the image will be reduced 4 times, compared to the infinity case.*

9.3 Ideal white light source with optical power of 1 W has uniform spectral density in the range 380 to 680 nm. Calculate the photometric output I in lumens.

Solution (9.3). *Using the approximation for spectral sensitivity Eq. 9.1 and conversion formula 9.2 we calculate*

$$I = 683 \int_{0.38}^{0.68} 1.019 \exp(-285.4(\lambda - 0.559)^2) \frac{1}{0.68 - 0.38} d\lambda \approx 243 \text{ lm}.$$

This result illustrates that the maximum achievable light output for a white light bulb is limited, and even with 100% conversion from electricity to light, the value of 683 lm/W is not achievable due to non-uniform spectral sensitivity of the eye. The realistic limit is somewhere in the range of 200 to 250 lm/W.

9.4 Archimedes equipped 100 soldiers with round shields of 1 m in diameter. Each shield has concave perfectly spherical surface with ROC of 300 m, made of polished bronze with reflectivity of 65%. All soldiers focus the Sun images in the same point on the sail of the enemy ship, at 150 m distance. Assume the sail reflectivity of 0.65

(the sails are rather dirty after the long campaign). Assume the solar irradiance of 1000 W/m^2, the angular diameter of Sun of 2000 arc seconds. It's a hot day with temperature of 30 C. Estimate the maximum possible temperature in the "united" focal spot, assuming all soldiers can point the Sun image to the same area. Compare the result to a more realistic estimate, when the soldiers are equipped with flat polished bronze shields of 1 m in diameter.

References

1. W.R. McCluney, *Introduction to Radiometry and Photometry*, 2nd edn. (Artech House, 2014)
2. F. Grum, *Radiometry* (Academic Press, 2012)
3. R.W. Boyd, *Radiometry and the Detection of Optical Radiation* (Wiley, 1983)
4. J.G. Graeme, *Photodiode Amplifiers: OP AMP Solutions* (McGraw-Hill Education, 1996)

Chapter 10
Introduction to Lasers

Abstract The laser is a source of coherent light, operating by stimulated emission within an active medium possessing a non-equilibrium energy level structure. Its operation is explained using three and four-level energy diagrams. The optical resonator, which defines the temporal and spatial properties such as mode structure and coherence of the output beam, is detailed. The fundamental relationships between the laser medium's properties, the resonator's geometry, and the resulting radiation parameters are established. Both stable and unstable resonator configurations are analyzed. Finally, various laser types are surveyed alongside their technological applications.

10.1 Lasers

For a long period in the history of optical engineering, no bright coherent sources were available. Light was produced only by thermal and luminescent sources. Thermal radiation has a very broad spectral range, which limits temporal coherence, while high spatial coherence requires a source with very small angular size.

The situation changed in the 1960s with the invention and application of lasers. The acronym LASER stands for *Light Amplification by Stimulated Emission of Radiation*. An authoritative and comprehensive review of laser science and technology is given by Siegman in [1]. Lasers provide unprecedented brightness and coherence, far exceeding those of natural sources.

In a laser, schematically shown in Fig. 10.1, coherent light is generated by combining:

- A resonant optical cavity, formed by two mirrors (M_1 and M_2), that filters the spatial and temporal modes of optical radiation to produce a highly coherent beam exiting through the partially transmitting mirror M_2, which simultaneously provides positive feedback to the gain medium.
- A resonant optical amplifier placed inside the cavity, formed by a medium with optical gain in a non-equilibrium thermodynamic state, where the population of higher quantum energy levels exceeds that of lower levels.

G. Vdovin, *Elementary Technical Optics*, UNITEXT for Physics,
https://doi.org/10.1007/978-3-032-08626-6_10

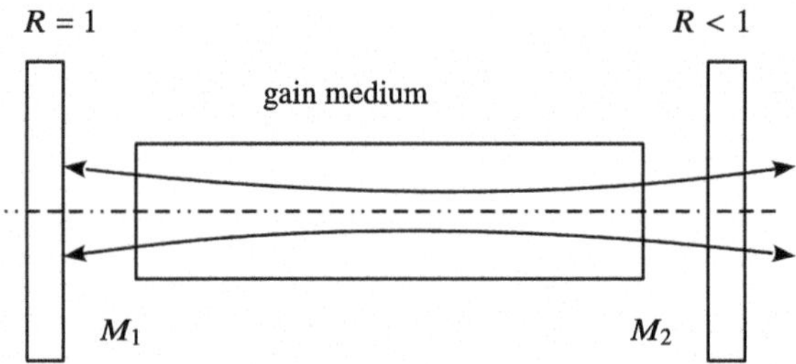

Fig. 10.1 Schematic diagram of laser operation

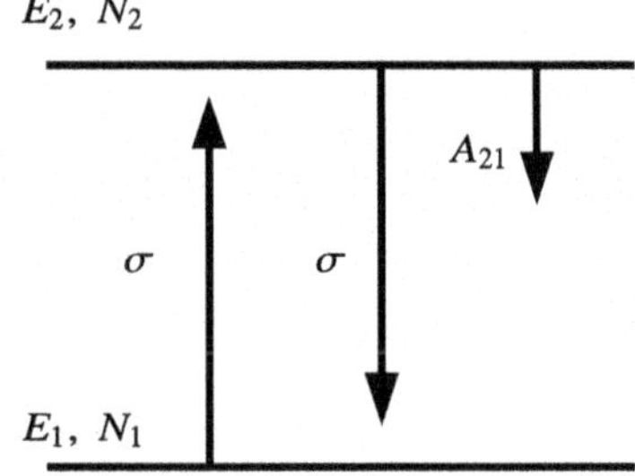

Fig. 10.2 Two-level system

10.2 Laser Energy Levels

In thermodynamic equilibrium, all media absorb light. The light intensity under absorption conditions decays exponentially with propagation distance z:

$$I_z = I_0 \mathrm{e}^{-\gamma z}, \tag{10.1}$$

where γ [m^{-1}] is the absorption coefficient. The coefficient γ depends on the concentration N [m^{-3}] of absorption centers (atoms, ions, molecules, or quantum dots) and on their effective cross section σ [m^2]:

$$\gamma = N\sigma. \tag{10.2}$$

Absorption can be described by a simple two-level system with E_1 and E_2 ($E_2 > E_1$), as shown in Fig. 10.2. The populations N_1 and N_2 of the two levels follow Fermi–Dirac statistics:

$$\frac{N_2}{g_2} = \frac{N_1}{g_1} \mathrm{e}^{-\frac{E_2 - E_1}{kT}}. \tag{10.3}$$

Since $E_2 > E_1$ and the levels are often non-degenerate ($g_1 = g_2 = 1$), it follows that $N_1 > N_2$.

Three types of quantum transitions are possible in such a system:

- *Stimulated absorption*: a photon is absorbed by a system in state E_1, raising it to state E_2.

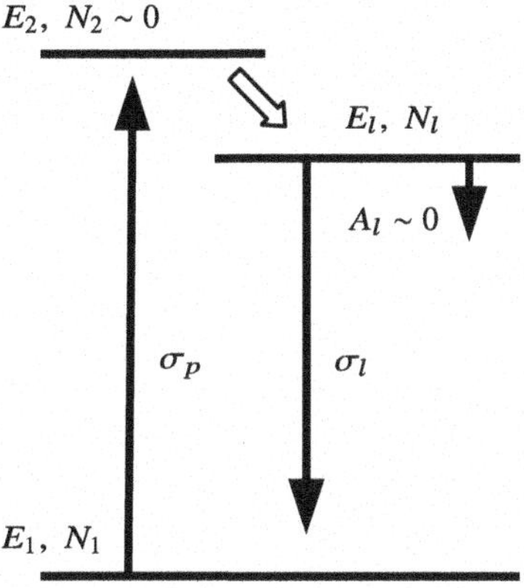

Fig. 10.3 Three-level system

- *Stimulated emission*: a system in state E_2 is induced by an incoming photon to emit a photon with the same phase and direction as the incoming one. This process provides optical amplification in lasers.
- *Spontaneous emission*: a system in state E_2 emits a photon spontaneously in a random direction.

Optical amplification requires stimulated emission to exceed stimulated absorption, which is equivalent to the condition

$$\Delta N = N_2 - N_1 > 0, \tag{10.4}$$

where ΔN is the *population inversion*, defining the gain according to 10.2. In a two-level system, this inversion cannot be achieved, since the population of E_2 can never exceed that of E_1, regardless of excitation strength.

Optical gain can be achieved in a luminescent medium under non-equilibrium conditions in the three-level system shown in Fig. 10.3. Here, level E_2 has a very short lifetime and is rapidly depopulated ($N_2 \sim 0$) through non-radiative transitions to the metastable laser level E_m. This level has a long lifetime and a very small spontaneous emission coefficient A_m. When the system is pumped by strong radiation Φ_p of wavelength λ_p matching the $E_1 \to E_2$ transition, a state with $N_m \gg N_1$ is created, leading to optical gain at the laser wavelength $\lambda_l = ch/(E_m - E_1)$.

The difference between the pump wavelength λ_p and the luminescence (laser) wavelength λ_l is the *Stokes shift*. Larger Stokes shifts correspond to higher thermal losses in the medium.

In a three-level system, positive gain requires $N_m > N_1$. Since $N_m + N_1 = N$, this condition implies $N_m > N/2$. Thus, more than half of the active centers must be excited to the metastable level, placing demanding requirements on the pump rate Φ_p.

The *four-level system*, shown in Fig. 10.4, avoids this limitation. The laser transition occurs between E_m and a very short-lived bottom level E_b, which is quickly

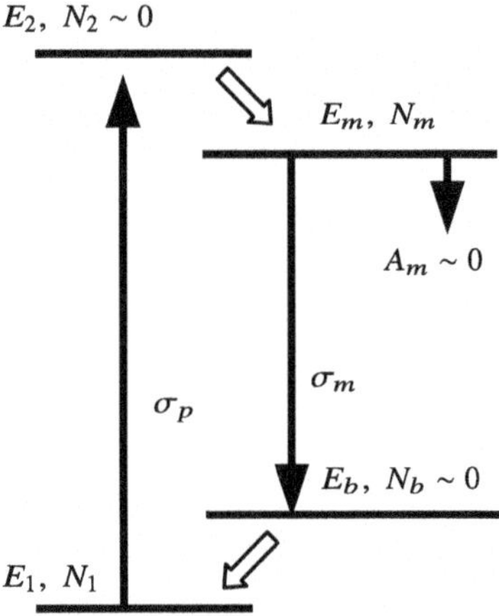

Fig. 10.4 Four-level system

depopulated into the ground state E_1 ($N_b \sim 0$). Thus, positive gain requires only $N_m > 0$, making lasing far easier to achieve than in three-level systems.

In special *self-limiting systems*, the bottom laser level depopulates slowly. In these lasers, the upper laser level is quickly populated by a powerful pump pulse, after which lasing continues until the population of the bottom level equals that of the upper level. Generation then stops until the bottom level is depopulated. Gold and copper vapor lasers, pumped by very fast electric discharge, operate in this regime.

At high intensities, the laser gain becomes saturated:

$$\gamma = \gamma_0 \frac{1}{1 + \Phi/\Phi_s}, \tag{10.5}$$

where $\gamma_0 = \sigma_m \Delta N$ is the small-signal (unsaturated) gain, Φ is the flux density, and Φ_s is the saturation flux density:

$$\Phi_s \approx \frac{h\nu}{\sigma_m \tau}, \tag{10.6}$$

with τ being the lifetime of the laser level.

Lasing in a cavity of length L (the resonator base), filled with a medium of gain γ_0 and saturation flux density Φ_s, is established if the total round-trip gain equals the feedback coefficient, approximately equal to the reflectivity R of the output mirror:

$$\exp\left(\frac{\gamma_0 2L}{1 + \Phi/\Phi_s}\right) = R, \tag{10.7}$$

with solution

$$\Phi = -\frac{\Phi_s(2\gamma_0 L + \ln R)}{\ln(R)}. \tag{10.8}$$

Equation 10.8 provides an engineering estimate of the flux density in the laser, assuming an idealized thin medium, sufficient pump rate, and neglecting spectral and spatial effects as well as diffraction losses. From 10.8, the obvious condition for laser generation follows:

$$R > \mathrm{e}^{-\gamma_0 2L}, \quad \gamma_0 = \sigma \Delta N. \tag{10.9}$$

In summary, lasing cannot be achieved in a two-level system because population inversion is impossible. In a three-level system, inversion is attainable but requires pumping more than half of the active centers, which imposes demanding pump requirements? In contrast, a four-level system requires only a small population of the upper laser level to exceed that of the quickly depopulated lower level, making it the most efficient and practical scheme for achieving optical gain.

10.3 Laser Resonators

The laser resonator provides feedback to the gain medium and acts as a resonant selector of spatial and temporal optical modes, thereby defining the output beam.

In the simplest case, the resonator consists of two spherical mirrors (with plane mirrors as a special case).

For analysis of spectral properties, it is convenient to use the Fabry–Perot model (Fig. 10.1), represented by two flat mirrors separated by distance L. The temporal selectivity of the resonator is determined by its structure of longitudinal modes. Each mode has a bandwidth $\delta\lambda$, and the modes are separated by

$$\Delta\nu = \frac{c}{2L}. \tag{10.10}$$

In a passive resonator (without gain), output coupling is provided by a mirror with reflectivity $R < 1$, while the second mirror is assumed fully reflective. With each round trip the photon number is multiplied by R. To find the number of round trips k at which the amplitude decays by e and the intensity by e^2, we set

$$R^k = \frac{1}{\mathrm{e}^2}; \quad k = -\frac{2}{\ln(R)}. \tag{10.11}$$

The coherence length of radiation inside the resonator equals $2Lk$, from which we estimate the resonance bandwidth $\delta\nu$:

$$\frac{\lambda^2}{\delta\lambda} = 2\,Lk, \quad \delta\lambda = -\frac{\lambda^2 \ln(R)}{4\,L}. \tag{10.12}$$

The spectral spacing between longitudinal modes is

$$\Delta\lambda = \frac{\lambda^2}{2L}. \tag{10.13}$$

Equations 10.12 and 10.13 define the longitudinal mode structure and spectral selectivity of the laser resonator. In practice, the realized resonance bandwidth is broader due to additional losses, mirror defects, misalignments, and line broadening in the gain medium.

The spatial properties of laser radiation are determined by the transverse modes of the resonator. The resonator can be treated as an infinite sequence of lenses forming a lens-like waveguide. Similar to optical fibers, such a waveguide supports a spectrum of eigenmodes, each described by a two-dimensional electromagnetic field distribution that reproduces itself after a round trip. Since each round trip involves diffraction on mirror edges, every mode experiences diffraction losses. If the gain is low, the mode with the lowest diffraction loss is generated first, just above threshold, yielding single-mode operation. At higher gain, multiple transverse modes satisfy the generation condition, and the laser operates in multimode regime. In this case, the beam exhibits reduced spatial coherence, higher divergence, and reduced temporal coherence.

In stable resonators, the field is confined by mirror curvature, with only a small fraction escaping by diffraction at the edges. Modes are selected solely by diffraction losses, so stable resonators with low diffraction losses provide poor mode selectivity. Resonators with longer bases L and smaller mirror apertures a have higher diffraction losses, thus better selectivity. By analogy to waveguides, longer resonators with smaller mirrors behave similar to single-mode waveguides.

An alternative is the unstable resonator, in which the cross-section of the fundamental mode expands geometrically after each round trip. Light is coupled out around the edges of the output mirror. If the geometrical magnification of the resonator is M, the feedback coefficient in geometric approximation is

$$F = M^{-2}, \tag{10.14}$$

and the output beam has a ring-shaped cross section. Unstable resonators are used in high-power lasers, since they offer strong transverse mode selectivity and are practical for large active medium diameters (Fig. 10.5).

Defining

$$g_1 = 1 - \frac{L}{R_1}; \quad g_2 = 1 - \frac{L}{R_2}, \tag{10.15}$$

where $R_{1,2}$ are the mirror radii of curvature and L is the cavity length, the resonator is stable if

$$0 < g_1 g_2 < 1. \tag{10.16}$$

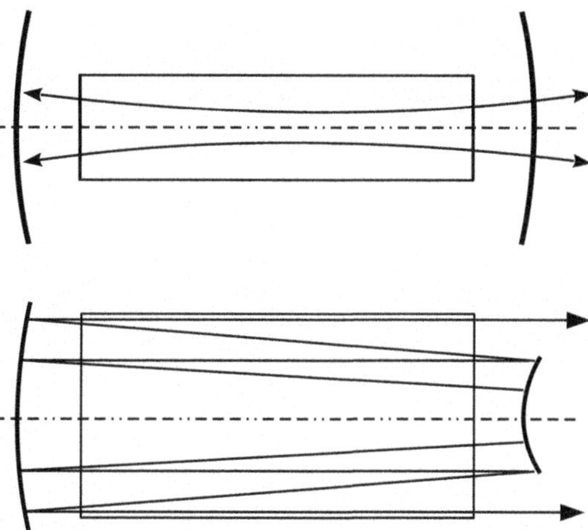

Fig. 10.5 Stable (top) and unstable (bottom) laser resonators

The Gaussian beam is the fundamental mode of a stable resonator. It is fully described by its complex radius of curvature $\widetilde{\rho}$ (Eq. 3.23) in any plane of the resonator. For a resonator mode, $\widetilde{\rho}$ repeats after a full round trip. The round trip is represented by the $ABCD$ matrix (Sect. 3.4), so the self-reproduction condition is

$$\widetilde{\rho} = \frac{A\widetilde{\rho} + B}{C\widetilde{\rho} + D}, \tag{10.17}$$

where $A \ldots D$ are elements of the round-trip matrix. Solving Eq. 10.17, we find the beam curvature ρ and waist w as functions of the round-trip matrix coefficients:

$$\rho = \frac{2B}{A - D}; \quad kw^2 = \frac{2|B|}{\sqrt{4 - (A + D)^2}}. \tag{10.18}$$

Since Eq. 10.18 gives beam parameters at any chosen plane (often taken at the output mirror), the waist position l_w is

$$l_w = \frac{D - A}{C}. \tag{10.19}$$

The divergence of laser beams is characterized by the M^2 parameter. If an ideal Gaussian beam has divergence φ and waist $w = \lambda/(\pi\varphi)$, while a real (multimode or aberrated) beam with the same waist has divergence $\varphi_r \geq \varphi$, then

$$\mathrm{M}^2 = \frac{\varphi_r}{\varphi}. \tag{10.20}$$

Single-mode lasers typically have $\mathrm{M}^2 = 1 \ldots 1.5$, whereas high-power multimode lasers may reach values up to 100 and higher.

10.4 Types of Lasers

Laser action has been demonstrated in hundreds of different materials, using electronic transitions in atoms and ions, rotational and vibrational transitions in molecules, excimer transitions, and quantum-dot transitions in solids, semiconductors, liquids, gases, and plasmas.

A three-level system is realized in ruby lasers (Al_2O_3 doped with Cr^+). Since such systems require very high pumping rates, ruby lasers typically operate in pulsed mode.

A four-level system is realized in Nd:YAG (yttrium–aluminum–garnet doped with Nd^+) and in He–Ne lasers. Because population inversion is achieved at very low pump rates, these lasers can operate in continuous-wave (CW) mode.

Excitation to the laser level can be achieved by various mechanisms. Optical pumping is used in solid-state and liquid lasers, driven by incandescent lamps, gas-discharge lamps, LEDs, or semiconductor lasers. Today, semiconductor lasers are preferred pump sources for solid-state lasers (yes, lasers are used to pump lasers!), as they deliver bright, narrowband radiation precisely matched to the absorption band of the gain medium.

Electrical pumping is widely used in semiconductor lasers and also in gas, metal-vapor, and excimer lasers.

Chemical pumping produces excited states through chemical reactions, though it is rarely applied due to issues with handling reaction products.

Gas-dynamic lasers (e.g., CO_2) employ rapid expansion and cooling of gas in specially shaped nozzles, creating a non-equilibrium state with population inversion in molecular levels. These systems resemble jet engines in operation.

With diverse gain media, laser emission has been achieved across the spectrum, from deep UV to far-IR and even into the terahertz range (masers). Laser beams can reach extremely high powers, exhibit exceptional spatial and temporal coherence, and generate ultrashort pulses down to the femtosecond and sub-femtosecond scale.

10.5 Applications of Lasers

Lasers are used across virtually all areas of modern science and technology.

Low-power, highly coherent He–Ne lasers are employed in metrology, optical testing, alignment, and lithography.

Semiconductor lasers are used as pump sources in solid-state and fiber lasers, in materials processing, and as general-purpose light sources in optics. They also dominate information technology, powering optical memory systems such as CD, DVD, and Blu-ray devices.

Solid-state lasers serve in materials processing, lidar, and medicine. Broadband solid-state media support generation of ultrashort (femtosecond) pulses for chemical and biological research and medical applications.

Gas lasers (e.g., CO_2) are widely used in manufacturing for cutting, welding, and surface processing.

From precise metrology to high-power industrial processing, lasers combine high coherence, brightness, and versatility, making them indispensable tools in science, technology, and industry.

10.6 Conclusion

Laser operation relies on population inversion in an active medium and resonant feedback in an optical cavity. Quantum systems with minimum three energy levels can be used for laser generation. However, three-level schemes require excitation of more than half of the active centers, imposing high requirement to the pumping. Four-level systems are the most practical, as inversion is achieved between the metastable upper laser and the lower laser level which is quickly emptied. Saturation effects in the gain medium limit the laser gain. The lasing threshold is determined by the balance between round-trip gain and cavity losses.

The resonator defines the temporal and spatial structure of the output beam. Longitudinal mode separation is determined by the cavity length. The spatial structure of the beam depends on the resonator geometry and the structure of transversal modes. In multimode operation, coherence length decreases proportionally to the number of oscillating longitudinal modes. Transverse modes are selected by diffraction in the resonator. Stable resonators confine fields by mirror curvature, with eigenmodes presented by low-loss Gaussian modes. Unstable resonators introduce geometric magnification, strongly discriminating against higher-order modes and enabling high-power operation with large gain volumes.

The Gaussian beam represents the fundamental mode of a stable resonator. Its waist position and size can be calculated by applying the $ABCD$ rule to the round trip matrix. The divergence and the beam quality are characterized by the M^2 parameter, providing a direct link between resonator geometry and measurable beam properties.

Different materials are used as active media in lasers. Ruby realizes a three-level system with pulsed operation, Nd:YAG and He–Ne represent efficient four-level systems with continuous emission. Excimer, solid-state and semicinductor lasers cover broad spectral range and can produce beams with excellent temporal and spatial characteristics. Pumping methods for different lasers include optical, electrical, and chemical excitation, suited to specific materials and applications.

10.7 Problems

10.1 Resonator of a He–Ne laser emitting at $\lambda = 633\,\text{nm}$ has base of $L = 1\,\text{m}$. The laser is generating 100 longitudinal modes. Estimate the coherence length L_c of the outgoing beam.

Solution (10.1). *Using 10.10 we obtain* $\Delta\nu \approx 100c/2L$, *then* $L_c \approx 2L/100$.

10.2 Resonator of a He–Ne laser emitting the fundamental mode at $\lambda = 633\,\text{nm}$. The resonator has base of $L = 1\,\text{m}$. The resonator is formed by flat output mirror and concave back mirror with radius of curvature of –4 m. Calculate the beam waist size, position, and the divergence of the output beam.

Solution (10.2). *Building the round trip matrix and using Eqs. 10.18 and 10.19 we obtain*

$$\rho = \infty,\ l_w = 0,\ w_0 = 448 \cdot 10^{-6}\text{m}$$

then the divergence $\varphi = \lambda/(\pi w_0) \approx 4.25 \cdot 10^{-4}$ *rad.*

10.3 Nd:YAG laser crystal has laser transition cross section of $\sigma = 2.8 \cdot 10^{-19}\,\text{cm}^2$. Active rod of a laser has total length of $L = 10\,\text{cm}$ and is placed into a laser resonator with semi-transparent output mirror having reflectivity of $R = 0.7$. Calculate the minimum concentration N of active centra in the rod, to provide laser generation.

Solution (10.3). *The total length of the gain medium in the round trip through the resonator equals* $2L = 20\,\text{cm}$. *Since the YAG laser utilizes 4-level scheme, the total concentration* N *of the active ions can be used to create the population inversion:* $\Delta N = N$. *Then, equating the total gain to the losses in the output, we obtain:*

$$\exp(2LN\sigma)R > 1,$$

with a solution

$$N > -\frac{\ln(R)}{2L\sigma} \approx 6.5 \cdot 10^{16}\text{cm}^{-3}$$

Reference

1. A.E. Siegman, *Lasers* (University Science Books, 1986)

Chapter 11
Introduction to Spectral Instruments

Abstract Spectral analysis is a powerful technique widely used to determine the chemical composition of materials, as each substance possesses an unique emission spectrum. These optical spectra are measured using specialized spectral instruments. A general spectral instrument consists of three key components: a sampling element (such as a slit), a dispersive element positioned at the pupil plane of the optical system, and an imaging system that projects the dispersed slit image onto an optical sensor. Prism spectrometers operate based on the dispersion of light in transparent optical materials. The resolving power of a prism spectrometer depends on the dispersion properties of the material and the physical size of the prism. Grating spectrometers utilize diffraction from periodic structures (gratings). For these instruments, the resolving power is shown to depend on the total number of illuminated grooves and the order of interference. Fabry-Perot spectrometers are used to achieve extremely high spectral resolution in a narrow spectral range.

11.1 Light Spectra

Spectral analysis is widely applied across science and technology because optical spectra provide detailed information about the unique structure of energy levels in matter. Analysis of spectral data reveals the composition of materials and their physical state.

The human eye is sensitive to only three regions of the visible spectrum, yet with this limited capability humans can distinguish millions of colors.

Emission spectra, unique for each element, arise from transitions between excited quantum energy levels in atoms, ions, and molecules. Absorption spectra, by contrast, appear when continuous-spectrum light is absorbed within specific energy bands characteristic of each material. Spectra from excited atoms and molecules in gases and plasmas are influenced by electric and magnetic fields as well as by particle motion. Spectroscopic study thus yields valuable information about composition, temperature, velocity, and other conditions in both laboratory and natural environments. For instance, the physical properties, composition, and dynamics of

G. Vdovin, *Elementary Technical Optics*, UNITEXT for Physics,
https://doi.org/10.1007/978-3-032-08626-6_11

celestial bodies, stars, and galaxies are determined primarily through their spectral signatures.

Spectral composition of light is measured with spectrometers. A good overview of the fundamentals of spectrometer design is provided in [1].

11.2 Basic Spectrometer

A general imaging spectrometer, shown in Fig. 11.1, consists of an entrance slit and an imaging system that forms an image of the slit in the detector plane. The position of the slit image in the detector plane is determined by the wavelength-dependent angular deflection introduced by the dispersive element, which is placed in the collimated beam between the imaging lenses. Typically, the dispersive element is positioned in the system pupil of diameter D. The source image is projected onto the entrance slit by an external imaging system. For example, a star image can be projected onto the slit by a telescope. To ensure maximum light throughput and the highest resolution, the exit pupil of this external system must be conjugated with the dispersive element via lens L_1, and the F-number of the external system must match that of L_1.

The resolving power of a spectral instrument is defined as

$$R = \frac{\lambda}{\Delta\lambda}, \tag{11.1}$$

where $\Delta\lambda$ is the minimum resolvable wavelength interval.

Since the spectrum is dispersed along the detector in the image plane of lens L_2, the minimum resolvable interval is defined by the half-width of the slit image and the angular dispersion of the dispersive element $d\varphi/d\lambda$, which determines the spectral broadening of the slit image:

$$\Delta y_s = \frac{d\varphi}{d\lambda}\Delta\lambda F_2 \tag{11.2}$$

in the focal plane of lens L_2.

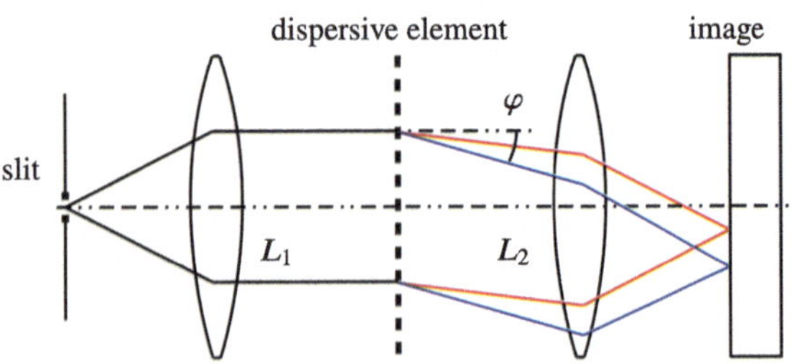

Fig. 11.1 Schematic of a general spectral instrument

For an infinitely narrow slit, the monochromatic image has a diffraction-limited half-width Δy_d:

$$\Delta y_d = \frac{\lambda F_2}{D}. \tag{11.3}$$

Equating 11.2 with 11.3, we obtain:

$$\frac{\lambda}{\Delta\lambda} = \frac{d\varphi}{d\lambda} D. \tag{11.4}$$

Thus, the theoretical resolving power of a spectral instrument depends only on the angular dispersion $d\varphi/d\lambda$ and the aperture D.

In summary, the resolving power of spectroscopic instruments is determined by the dispersive properties of the element and the aperture of the system.

11.3 Prism Spectral Instruments

The resolving power of a prism spectrometer depends on both the material dispersion and the prism aperture D. In the visible range, glasses with low Abbe number (e.g., heavy flints with high dispersion) are preferred. Near-UV applications employ crystalline quartz, fused silica, LiF, and CaF_2. In the far-IR, common materials include NaCl and KCl (up to $\lambda = 16\ \mu m$), KBr (20 μm), and CsI (up to 32 μm).

The angular dispersion of the prism (Fig. 11.2) is

$$\frac{d\varphi}{d\lambda} = \frac{\partial\varphi}{\partial n}\frac{dn}{d\lambda}, \tag{11.5}$$

where $\frac{dn}{d\lambda}$ is the material dispersion, and $\frac{\partial\varphi}{\partial n}$ depends on the prism parameters (apex angle θ, incidence angle ϕ_1):

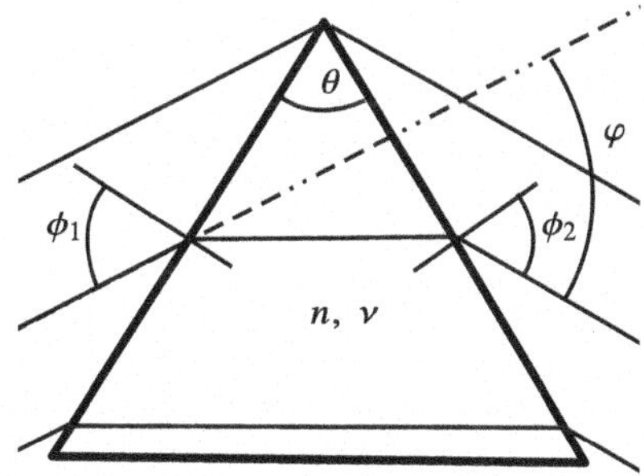

Fig. 11.2 Prism at minimum deviation

$$\frac{\partial\varphi}{\partial n} = \frac{\sin\theta}{\cos\phi_1\cos\phi_2}. \tag{11.6}$$

At minimum deviation, where the ray path through the prism is symmetric,

$$\left.\frac{\partial\varphi}{\partial n}\right|_{min\ dev} = \frac{2\sin\frac{\theta}{2}}{\sqrt{1-n^2\sin^2\frac{\theta}{2}}}. \tag{11.7}$$

Thus, $\frac{\partial\varphi}{\partial n}$ also depends on ϕ_1, being larger for smaller incidence angles. In practice, the prism design is determined not only by dispersion but also by the beam aperture and the meridional magnification due to cross-section changes inside the prism.

The total dispersive angle of the beam is estimated as

$$\Delta\varphi = \frac{\partial\varphi}{\partial n}\Delta n, \tag{11.8}$$

where Δn corresponds to the material's dispersion across the working wavelength range $\Delta\lambda$. In the simplest approximation, using linear interpolation and the definition of the Abbe number from 4.22:

$$\frac{dn}{d\lambda} \approx \frac{n-1}{\nu(\lambda_F-\lambda_C)}, \tag{11.9}$$

with $\lambda_F = 486.1$ nm, $\lambda_C = 656.3$ nm (Fraunhofer lines), and ν the Abbe number.

Combining 11.7 and 11.9 gives the relation between the wavelength interval and the dispersion angle for minimum deviation geometry:

$$\Delta\varphi \approx \frac{n-1}{\nu(\lambda_F-\lambda_C)}\frac{2\sin\frac{\theta}{2}}{\sqrt{1-n^2\sin^2\frac{\theta}{2}}}\Delta\lambda. \tag{11.10}$$

Hence, the angular dispersion is

$$\frac{d\varphi}{d\lambda} \approx \frac{n-1}{\nu(\lambda_F-\lambda_C)}\frac{2\sin\frac{\theta}{2}}{\sqrt{1-n^2\sin^2\frac{\theta}{2}}} \approx \left.\frac{n-1}{\nu(\lambda_F-\lambda_C)}\theta\ \right|_{\theta\ll 1}, \tag{11.11}$$

valid for $\lambda_F \leq \lambda \leq \lambda_C$.

The total angular dispersion across this interval is therefore

$$\varphi_C - \varphi_F \approx \left.\frac{(n-1)\theta}{\nu}\ \right|_{\theta\ll 1}. \tag{11.12}$$

Finally, the resolving power at the central wavelength λ_D is estimated as:

$$R = \frac{d\varphi}{d\lambda} D \approx \frac{n-1}{\nu(\lambda_C - \lambda_F)} \theta D \Bigg|_{\theta \ll 1}. \tag{11.13}$$

In this approximation, the resolving power increases with beam diameter D, decreases with Abbe number ν, and grows with prism apex angle θ, although the actual dependence $R(\theta)$ for large θ is nonlinear.

Prism spectrometers, while limited by the material dispersion and Abbe number, remain useful across wide spectral ranges by choosing appropriate materials. The performance scales with aperture size and prism geometry, though practical designs must balance dispersion, transmission, and beam size.

11.4 Grating Instruments

Most modern spectral instruments employ reflective diffraction gratings. Reflection from a flat diffraction grating with period d is illustrated in Fig. 11.3. The wavelength-dependent diffraction law for a reflective grating is given by:

$$\sin\phi + \sin\varphi = \frac{k\lambda}{d}, \tag{11.14}$$

where k is the diffraction order (spectral order) and d is the grating period. Angles are considered positive when measured counterclockwise from the normal, so in Fig. 11.3 $\phi < 0$ and $\varphi > 0$.

In the configuration shown in Fig. 11.3, the direction of specular reflection ($k = 0$) differs from that of the diffracted orders, and the efficiency of diffraction into higher

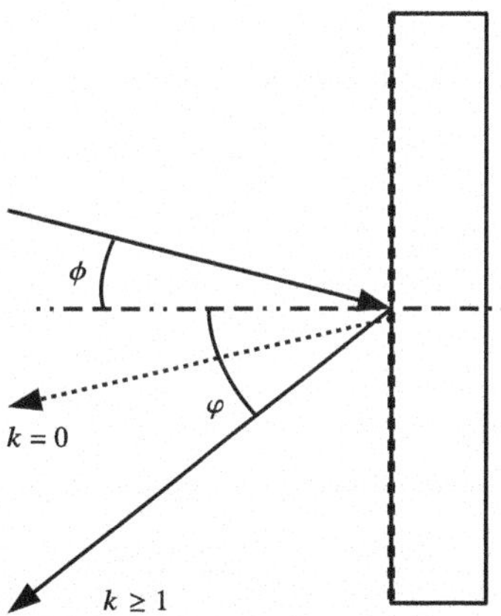

Fig. 11.3 Reflection from a diffraction grating

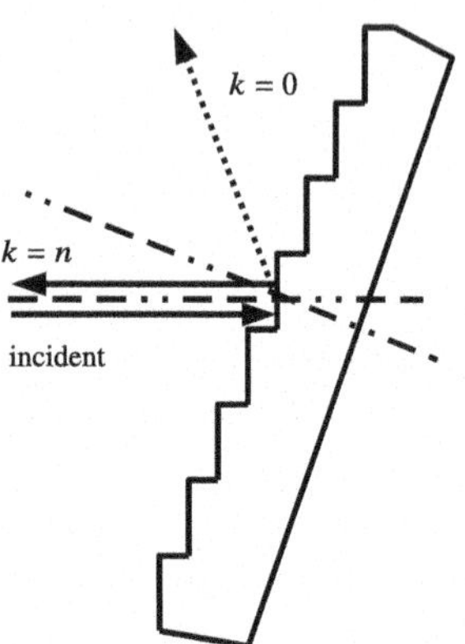

Fig. 11.4 Reflection from a blazed grating in autocollimation configuration

orders is relatively low. The efficiency can be increased by shaping the grooves so that specular reflection from the groove facets (the so-called blaze) coincides with the chosen diffraction order, as illustrated in Fig. 11.4.

Gratings can be manufactured with blaze angles optimized for very high diffraction orders. Gratings with relatively large groove spacing d, a small total number of grooves, and operation at high orders k are known as *echelle gratings*. To avoid overlapping of spectral orders, an order-separating element (typically a prism oriented orthogonally to the grating dispersion) is used.

The fabrication of gratings requires extremely precise control of groove periodicity. Imperfections or non-uniformities in groove spacing lead to the appearance of *ghost spectra*.

The angular dispersion of a grating is obtained by differentiating Eq. 11.14 with respect to wavelength λ:

$$\frac{d\varphi}{d\lambda} = \frac{k}{d\cos\phi} - \frac{\sin\phi}{\cos^3\phi}\frac{k^2\lambda}{d^2} + O(\lambda^2). \tag{11.15}$$

To first order, the grating dispersion is proportional to the diffraction order k and inversely proportional to the grating period d and $\cos\phi$. The second term in Eq. 11.15 introduces a wavelength dependence, which becomes significant at high orders.

The resolving power of a grating spectrometer can be expressed using Eq. 11.4:

$$R = \frac{d\varphi}{d\lambda}D = \frac{kD}{d\cos\phi} \approx \frac{kN}{\cos\phi}, \tag{11.16}$$

where $N = D/d$ is the total number of grooves illuminated on the grating.

Thus, the resolving power of a grating spectrometer with normal incidence on the grating is directly proportional to the number of illuminated grooves and the diffraction order. Flat reflective gratings are available with different blaze angles, optimized for particular wavelength ranges.

11.5 Monochromators

A monochromator isolates a narrow spectral band from a broader spectrum. A typical spectrometer can be converted to a monochromator by introducing the exit slit in the observation plane. The slit isolates a narrow spectral band from the whole observed spectrum. Obviously, the minimum spectral band $\Delta\lambda$ that can be isolated by a monochromator is

$$\Delta\lambda \geq \frac{\lambda}{R},$$

where R is the resolving power of the spectral instrument.

The output wavelength λ is usually adjusted by rotating the dispersive element, and/or moving the slit. Then a divergent beam in the wavelength range $\lambda \ldots \lambda + \Delta\lambda$ can be produced in the exit slit. Monochromators are frequently used to feed optical instruments such as Fabry-Perot interferometer with incoherent quasi-monochromatic light, therefore the numerical aperture of the receiving instrument should be matched to the numerical aperture of the monochromator exit beam, to avoid light losses.

11.6 Fabry-Perot Interferometer

The resolving power of grating and prism instruments is insufficient to distinguish the fine structure of spectral lines. A Fabry-Perot interferometer [2], consisting of two parallel semi-transparent mirrors, can achieve extremely high resolving power R, but only within a narrow spectral range $\Delta\lambda$. A typical configuration of a Fabry-Perot interferometer, coupled to the exit slit of a monochromator, is shown in Fig. 11.5. The interferometer is formed by two semi-transparent mirrors of reflectivity r, separated by distance b.

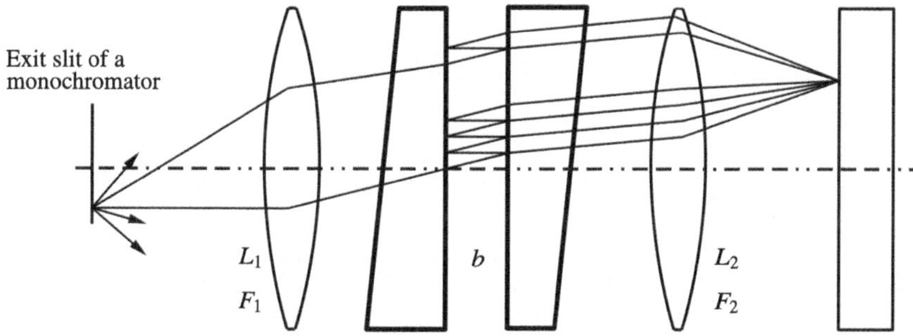

Fig. 11.5 Fabry-Perot interferometer used to resolve the structure of a spectral line

Each point of the slit produces a diverging beam in a limited spectral range $\lambda \ldots \lambda + \Delta\lambda$. This beam is collimated by lens L_1, enters the interferometer, undergoes multiple reflections between the mirrors, and is then focused by lens L_2 onto the screen. Multiple reflections inside the interferometer give rise to multi-beam interference in the observation plane. If the incident amplitude is U_0, the amplitude reflectivity of the mirrors is r, and the phase difference after a round trip of length $2b$ is $4\pi b/\lambda + \delta$, then the output amplitude U is the sum of a geometric progression:

$$U = U_0(1 - r)[1 + r\mathrm{e}^{i\delta} + r\mathrm{e}^{i2\delta} + \ldots],$$

which, according to A.1 evaluates to

$$U = U_0 \frac{1 - r}{1 - r\mathrm{e}^{i\delta}}.$$

The output intensity is

$$I(\delta) = |U|^2 = \frac{I_0}{1 + 4r\sin^2(\delta/2)/(1 - r)^2}. \tag{11.17}$$

Thus, the intensity depends on the mirror reflectivity r and the phase shift δ accumulated during the round trip.

Equation 11.17 describes a structure of very sharp fringes with long gaps in between. The positions of these fringes depend on the wavelength. To estimate the fringe width, and then the instrument resolving power, in terms of δ in the denominator of 11.17, note that the fringe intensity falls to half of the maximum when

$$F \cdot \sin^2(\delta/2) = 1,$$

where F is defined as

$$F = \frac{4r}{(1 - r)^2}.$$

Assuming $\sin^2(\delta/2) \approx \delta^2/4$, we obtain

$$\delta \approx \frac{1 - r}{\sqrt{r}} + 2\pi \frac{2b}{\lambda},$$

where $2b/\lambda = m$ is the interference order. The phase sensitivity of δ is therefore realized over the total phase $2\pi m = 4\pi b/\lambda$. The resolving power of the Fabry-Perot interferometer can then be estimated as

$$R \approx \frac{4\pi b\sqrt{r}}{(1 - r)\lambda}, \tag{11.18}$$

with the minimum resolvable spectral interval

$$\delta\lambda = \frac{\lambda}{R} = \frac{(1-r)\lambda^2}{4\pi b\sqrt{r}}. \tag{11.19}$$

To determine the spectral spacing between fringes, note that constructive interference requires $\delta = 2\pi m/\lambda$, where m is the order of interference. For adjacent fringes, $\Delta\lambda$ satisfies

$$m(\lambda + \Delta\lambda) = (m+1)\lambda,$$

with the solution

$$\Delta\lambda = \frac{\lambda}{m} \approx \frac{\lambda^2}{2b}. \tag{11.20}$$

This condition defines the required monochromator bandwidth feeding the Fabry-Perot interferometer. If the spectral range of the feeding instrument is broader, several overlapping fringe systems will appear. Also, since different angles θ correspond to different spectral components, the feeding source for the Fabry-Perot spectrometer should have sufficient angular size to secure all necessary parts of the fringe pattern are imaged in the output.

In conclusion, the Fabry-Perot interferometer provides exceptionally high spectral resolution by exploiting multi-beam interference formed by reflections between highly reflective mirrors. Its resolving power grows with mirror reflectivity and spacing, but its operational spectral range is narrow, therefore it is used coupled to a filter or a monochromator. Then the Fabry-Perot interferometer enables the observation of fine spectral details that cannot be resolved with grating or prism instruments. Fabrication of Fabry-Perot etalons requires extreme precision in mirror polishing and alignment. Also semi-transparent mirror coatings should satisfy to very high requirements.

11.7 Comparison of Spectral Instruments

Spectral instruments based on reflective gratings combined with mirror optics offer several advantages:

- Broad usable spectral range, from deep UV to far IR.
- High dispersion that, to first order, is independent of wavelength.
- Resolving power R determined by the number of illuminated grooves, independent of the physical aperture size.
- Gratings can be manufactured with high reflectivity and large apertures, whereas in prisms, absorption in highly dispersive materials becomes significant at large sizes needed for high resolution.

Advantages of prism instruments include:

- Spectra free from overlapping diffraction orders and ghosting effects.
- More uniform transmission losses across the spectrum, whereas grating reflectivity can vary strongly with wavelength, especially when the blaze angle is mismatched to the diffraction order.
- Broad spectral coverage is straightforward with prisms, while gratings achieve about one octave of spectral range only in the first order.

Fabry-Perot interferometers are used to resolve fine structure of spectral lines, usually in combination with a narrowband spectral filter or monochromator.

In summary, grating instruments are generally preferred when very high spectral resolution, wide wavelength coverage, or compact instrument design is required, while prisms remain advantageous in applications that demand clean spectra without overlapping orders, and stable transmission across the full visible range. The choice between a grating and a prism thus depends on the trade-off between spectral purity and resolution. Fabry-Perot instruments are used to resolve fine structure of spectral lines.

11.8 Conclusion

Spectral instrument consist of an entrance slit, a dispersive element placed in the pupil, and an imaging system projecting the dispersed slit image onto a detector. The resolving power is set by the angular dispersion of the element and the diffraction on system aperture.

In prism spectrometers, dispersion is defined by glass properties. The achievable resolution increases with the prism aperture, apex angle, and material dispersion. Heavy flints provide high dispersion in the visible, while quartz, CaF_2, and LiF are used in the UV. Halide crystals extend operation into the IR. Resolution scales with aperture, though large prisms introduce higher absorption losses.

In grating spectrometers, dispersion is defined by interference at periodic structures. The resolving power depends on the diffraction order and the number of illuminated grooves. Echelle gratings operate at high diffraction orders co-incident with specular reflection. Fabrication precision is critical for gratings, since periodicity errors introduce ghost spectra.

Prism instruments provide continuous spectra free from overlapping orders and wavelength-dependent efficiency. Grating instruments achieve higher resolution and broader spectral coverage, but can suffer from overlapping orders.

The Fabry-Perot interferometer provides exceptionally high spectral resolution by exploiting multi-beam interference between highly reflective parallel mirrors. Its resolving power grows with mirror reflectivity and spacing, but in a narrow wavelength range. The requirements to the precision of fabrication and alignment of the components of a Fabry-Perot interferometer are exceptionally high.

11.9 Problems

11.1 Show that the resolving power of prism spectral instrument, shown in Fig. 11.6, is proportional to the glass dispersion and the size of the prism base T:

$$R \approx \frac{dn}{d\lambda} T.$$

Solution (11.1) Assuming $\theta \approx T/A$ and using the paraxial approximation, we have

$$\varphi \approx (n-1)\frac{T}{A}.$$

The resolution $\Delta\lambda$ can be obtained from

$$\frac{d\varphi}{d\lambda}\Delta\lambda = \frac{\lambda}{A},$$

where

$$\frac{d\varphi}{d\lambda} = \frac{dn}{d\lambda}\frac{T}{A}.$$

Equating

$$\frac{dn}{d\lambda}\frac{T}{A}\Delta\lambda = \frac{\lambda}{A}.$$

we obtain

$$R = \frac{\lambda}{\Delta\lambda} = \frac{dn}{d\lambda} T.$$

11.2 Prism spectrometer uses dispersion prism fabricated from glass with MIL index 600300, with aperture $D = 2$ cm, and apex angle of $\theta = 0.1$ rad. Estimate the resolving power R at the wavelength λ_D.

Solution (11.2) Using Eq. 11.16 we directly calculate:

$$R \approx \frac{n-1}{\nu(\lambda_C - \lambda_F)}\theta D \approx 235$$

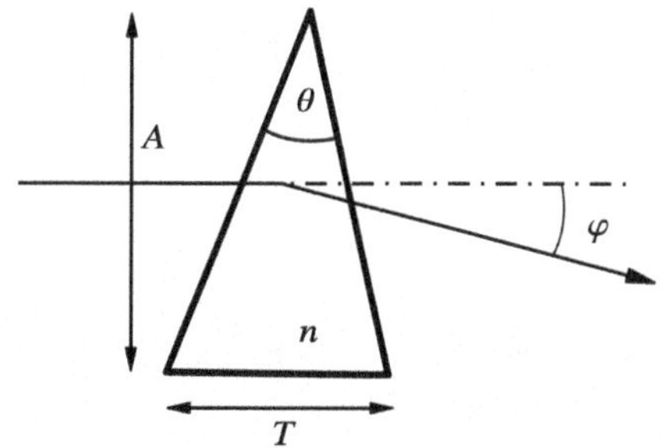

Fig. 11.6 Illustration to Problem 11.1

11.3 Echelle grating working in the autocollimation configuration in the 5-th order of interference $k = 10$ has the groove width of $d = 10\ \mu$m and the blaze angle of $\phi = \varphi = 20°$. Calculate the wavelength λ at which the spectrometer has its maximum efficiency.

Solution (11.3) Using Eq. 11.14 we can equate

$$\sin\phi + \sin\varphi = \frac{k\lambda}{d},$$

then

$$\lambda = \frac{\sin\phi + \sin\varphi}{kd} \approx 6.87 \cdot 10^{-7}\ \text{m}$$

11.4 Design Fabry-Perot interferometer to resolve sodium yellow doublet of $\lambda_1 =$ 589 nm, and $\lambda_2 = 589.6$ nm

Solution (11.4) Assuming the free spectral range to be $\Delta\lambda = 10$ nm, and the required spectral resolution $\delta\lambda = 0.1$ nm, using expressions (11.19) and (11.20), we obtain for the distance between mirrors $b \approx 35 \cdot 10^{-6}$ m, and for the mirror reflectivity $r \geq$ 0.88. This example demonstrates that pretty decent resolving power can be achieved with modest parameters. Practical Fabry-Perot instruments achieve resolving powers of several magnitudes higher.

References

1. J. James, *Spectrograph Design Fundamentals* (Cambridge University Press, 2007)
2. J.M. Vaughan, *The Fabry-Perot Interferometer: History, Theory* (CRC Press, Practice and Applications, 2017)

Chapter 12
Introduction to Optics of Waveguides

Abstract This section describes the fundamental principles of optical waveguides. The primary design parameters for both step-index and graded-index waveguides are introduced, and the relationship between these parameters and key waveguide properties–such as the numerical aperture and the number of supported modes–is established. Coupling efficiency into waveguides is analyzed using the overlap integral. Finally, the bandwidth of the transferred optical signal is derived as a function of both material dispersion and the waveguide's numerical aperture.

12.1 Practical Reasons to Use Waveguides

The power density of a light beam generated by a source in free space decreases with distance Z as Z^{-2}. Even for very bright sources, transmitting optical power or optical signals over long distances leads to major power losses. In addition, atmospheric turbulence distorts optical wavefronts, so free-space transmission of optical information along narrow beams is limited to relatively short distances. High-speed free-space optical communication in the Earth's atmosphere is typically restricted to only a few kilometers, since longer distances result in a significant decrease of the signal-to-noise ratio and the transmission rate. Finally, the curvature of the Earth itself limits line-of-sight transmission.

The ability to guide light waves over long distances is therefore of great importance both for applications requiring the transfer of optical energy and for communication systems using light as an information carrier. The wide frequency range of optical signals enables very high information transfer rates, offering a clear advantage over wired and radio-frequency transmission systems.

G. Vdovin, *Elementary Technical Optics*, UNITEXT for Physics,
https://doi.org/10.1007/978-3-032-08626-6_12

12.2 Slab and Rectangular Waveguides

Consider a transparent optical slab with refractive index n_1, surrounded by material with refractive index n_2. A ray enters the slab from a medium with refractive index n_0 at an incidence angle φ with respect to the slab axis. If the condition $n_2 < n_1$ is satisfied, then rays propagating at angle φ_1 to the slab axis will undergo total internal reflection at the interface. Geometrical analysis (see Fig. 12.1) shows that

$$n_0 \sin\varphi = n_1 \sin\varphi_1 = n_1 \sin(\pi - \phi) = n_1 \cos\phi.$$

At the critical angle of total internal reflection, $\sin\phi = n_2/n_1$. Substituting $\cos\phi = \sqrt{1-\sin^2\phi}$ gives

$$n_0 \sin\varphi = n_1\sqrt{1-\left(\frac{n_2}{n_1}\right)^2} = \sqrt{n_1^2 - n_2^2}. \tag{12.1}$$

Expression 12.1 defines the numerical aperture A of the slab waveguide. Any ray with incidence angle smaller than $\varphi = \arcsin\frac{\sqrt{n_1^2-n_2^2}}{n_0}$ will be confined by total internal reflection inside the slab. It will propagate by bouncing between the slab interfaces. For a slab with small numerical aperture $\varphi \ll 1$ in air ($n_0 = 1$), we can approximate $A \approx \varphi \approx \sqrt{n_1^2 - n_2^2}$.

The waveguide input with size $2a$ limits the physical aperture for light coupling. To satisfy the condition of total reflection, the diffraction divergence of a plane wave entering the waveguide must not exceed the maximum acceptance angle φ (see Fig. 12.1). The divergence of a collimated beam with half-width a_{min} equals $\lambda/(2a_{min})$. Equating this divergence to the numerical aperture gives the minimum slab half-thickness a_{min}:

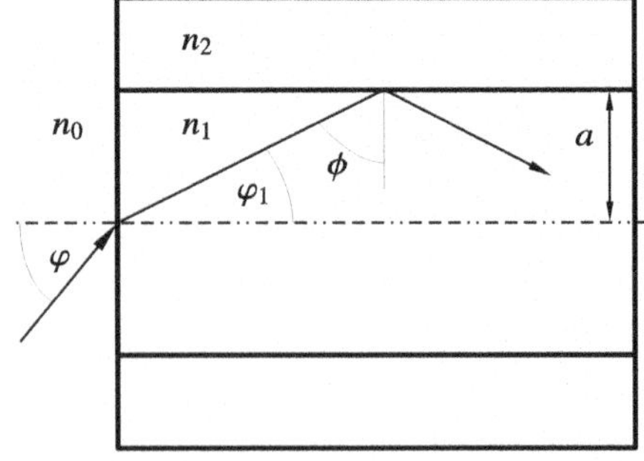

Fig. 12.1 Geometry of a slab waveguide

$$\frac{\lambda}{2a_{min}} = \sqrt{n_1^2 - n_2^2},$$

$$a_{min} = \frac{\lambda}{2A} \approx \frac{\lambda}{2\sqrt{n_1^2 - n_2^2}}. \quad (12.2)$$

Equation 12.2 defines the minimum slab size that supports guided propagation with only a single mode. If the slab thickness is smaller than $2a_{min}$, the wave decays due to leakage into the cladding. Conversely, if the slab thickness $2a$ is much larger than $2a_{min}$, the waveguide can support multiple modes. The approximate number of modes N is

$$N \approx \frac{a}{a_{min}} = \frac{2a}{\lambda}\sqrt{n_1^2 - n_2^2}. \quad (12.3)$$

The number of supported modes can also be derived by considering interference of plane waves propagating at angle φ_1. After reflection, the waves interfere periodically, forming modes analogous to diffraction patterns with period p. In the paraxial approximation, the minimum supported period p_{min} satisfies

$$\frac{A}{n_1} p_{min} = N\frac{\lambda}{n_1}. \quad (12.4)$$

The minimum waveguide size supporting at least one mode corresponds to $N = 1$, giving $p_{min}A = \lambda$ and

$$a_{min} = p_{min}/2 \approx \frac{\lambda}{2A} = \frac{\lambda}{2\sqrt{n_1^2 - n_2^2}}. \quad (12.5)$$

The number of modes N is then

$$N = \frac{pA}{\lambda} = \frac{2a}{\lambda}\sqrt{n_1^2 - n_2^2}. \quad (12.6)$$

For a rectangular waveguide of dimensions $a \times b$, the total number of modes N_{ab} is the product of the mode numbers for each dimension:

$$N_{ab} = N_a N_b = \frac{4ab}{\lambda^2}(n_1^2 - n_2^2). \quad (12.7)$$

Expression 12.7 gives the number of modes for a single polarization state. The total number of modes for both polarizations is $2N_{ab}$.

12.3 Step-Index Waveguides with Round Cross Section

Step-index waveguides with round cross section consist of a cylindrical core of radius a with refractive index n_1, surrounded by cladding with refractive index n_2.

To estimate the number of modes in a round step-index waveguide, we assume a Gaussian beam represents the fundamental mode. For a single-mode waveguide with core diameter $d = 2a_{min}$, we equate the Gaussian beam divergence to the numerical aperture:

$$\frac{\lambda}{\pi a_{min}} = \sqrt{n_1^2 - n_2^2},$$

$$a_{min} = \frac{\lambda}{\pi\sqrt{n_1^2 - n_2^2}}. \tag{12.8}$$

Expression 12.8 provides a good estimate of the minimum single-mode diameter. In engineering practice, optimal confinement of about 90% of the optical power within the core is reached at

$$a \approx 1.2a_{min} = 1.2\frac{\lambda}{\pi\sqrt{n_1^2 - n_2^2}}.$$

The effective area occupied by a single mode in one polarization state equals πa_{min}^2. Thus, for a multimode waveguide with core radius a, the number of modes per polarization state can be estimated as

$$N_p = \left(\frac{a}{a_{min}}\right)^2 = \left(a\frac{\pi}{\lambda}\sqrt{n_1^2 - n_2^2}\right)^2. \tag{12.9}$$

In practice, waveguides are characterized by the *V-number*:

$$V = a\frac{2\pi}{\lambda}\sqrt{n_1^2 - n_2^2} = akA, \tag{12.10}$$

where k is the free-space wavenumber and A is the numerical aperture. The V-parameter can be interpreted as a wavelength-scaled analogue of the Lagrange invariant, since it contains the product of the fiber's physical aperture and its numerical aperture. Just as the Lagrange invariant defines the number of resolvable elements in an optical system (see Eq. 3.12), the V-parameter determines the number of one-dimensional waveguide modes.

The number of two-dimensional modes supported by a circular fiber is $V^2/4$ per polarization state. Since each mode can exist in two independent polarization states, the total number of modes in a step-index fiber with round core is

$$N = 2N_p = \frac{V^2}{2}. \tag{12.11}$$

A round step-index fiber is considered *single-mode* when $V \leq 2.405$.

12.4 Coupling to Waveguides

Light is typically coupled into a waveguide using a lens that projects an image of the light source onto the waveguide input face. Figure 12.2 illustrates the coupling of a diffuse source into a waveguide. Input rays may either be guided into the core or leak into the cladding. The fraction of leaking rays depends on the numerical aperture of the source, φ, which is defined by the focusing lens, and the radius of the source image on the waveguide face, ρ. Efficient coupling requires minimizing the fraction of leaking rays. Optimal coupling of an incoherent source occurs under the conditions:

$$\varphi \approx \sqrt{2}A, \tag{12.12}$$

$$\rho \approx \sqrt{2}a, \tag{12.13}$$

where A is the waveguide numerical aperture and a is the core radius.

For coupling into a single mode, the efficiency is described by the *overlap integral*:

$$T = \frac{\left[\int_S E_s E_m^* ds\right]^2}{\int_S E_m E_m^* ds \int_S E_s E_s^* ds}, \tag{12.14}$$

where E_s and E_m are the fields of the source and waveguide mode, respectively, and $*$ denotes complex conjugation. Clearly, $T = 1$ if $E_s = E_m$.

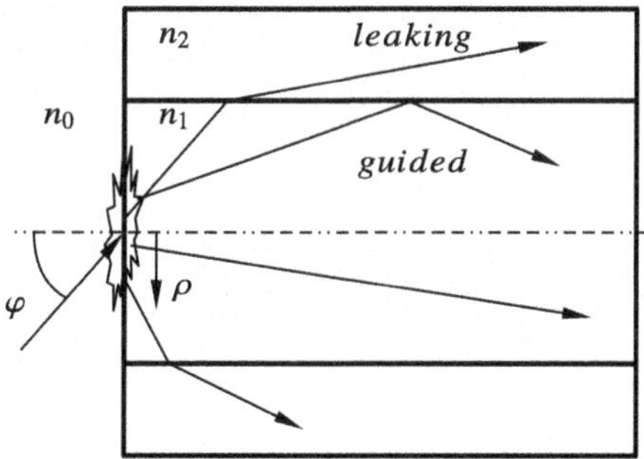

Fig. 12.2 Guided and leaking rays at the fiber input

The power of leaky rays not supported by the waveguide decays by 90% after a distance

$$Z = \frac{n_1 a}{2A} \exp(V/2). \tag{12.15}$$

A stationary modal distribution is established only at distances $L > Z$. The decay length for higher-order modes grows exponentially with the number of supported modes, making it very long in multimode waveguides. Consequently, a multimode waveguide with $L < Z$ behaves like a scattering element, with the output field strongly dependent on input conditions and mechanical bending. A stationary multimode waveguide produces a speckle field at its output. Temporal averaging of speckle coherence can be achieved by non periodic bending or distortion of the waveguide.

According to Eq. 12.15, the field distribution in a single-mode waveguide with $V \sim 2.4$ stabilizes over a relatively short distance. Single-mode fibers with $L > Z$ can therefore be used as mode filters, where the output field is weakly dependent on the input. However, as V increases into the multimode regime, filtering properties are rapidly lost due to the exponential growth in Eq. 12.15.

Waveguide bends and internal defects perturb the modal distribution, introducing additional leaky modes and altering the balance among guided modes. Hence, practical limits exist for bending radius and defect density in waveguides.

12.5 Bandwidth Limitations

A single-frequency signal carries no information; information transfer requires broadband modulation. Suppose a waveguide carries pulses of duration τ, corresponding to signal bandwidth $\Delta\nu = 1/\tau$. This bandwidth corresponds to a wavelength interval:

$$\Delta\lambda = \frac{\lambda^2}{\tau c}. \tag{12.16}$$

If the pulse propagates in a medium with refractive index n and dispersion $dn/d\lambda$, the index variation over $\Delta\lambda$ is

$$\Delta n = \Delta\lambda \frac{dn}{d\lambda}, \tag{12.17}$$

where λ is the central wavelength. Over a distance L, the total group delay is

$$T = \frac{nL}{c}.$$

The difference in arrival times between spectral components at λ and $\lambda + \Delta\lambda$ is

$$\Delta T = \frac{\Delta n L}{c} = \frac{nL}{c}\frac{\lambda^2}{\tau c}\frac{dn}{d\lambda}.$$

For distortion-free transmission, this delay must be much smaller than the pulse duration ($\Delta T \ll \tau$), yielding

$$L \ll \frac{\tau^2 c^2}{n\lambda^2 (dn/d\lambda)}. \tag{12.18}$$

Applying Eq. 12.18 to a picosecond pulse ($\tau = 10^{-12}$ s) at $\lambda = 633$ nm in BK7 glass gives a maximum undistorted propagation distance $L \ll 6.4$ m. Modern single-mode communication fibers are made from specialty glasses with near-zero dispersion and low absorption at the operating wavelength, allowing optical links of hundreds of kilometers without intermediate amplifiers or repeaters.

Equation 12.18 accounts for material dispersion in a single-mode waveguide, where the mode propagates along the axis. In multimode fibers, modes propagate at different angles within the numerical aperture A. For a fiber of length L, the propagation time difference between the axial mode and the mode at maximum angle $\varphi_1 = A/n$ is

$$\Delta T = \left(\sqrt{1 + (A/n)^2} - 1\right)\frac{Ln}{c} \approx \frac{A^2 L}{2nc}, \tag{12.19}$$

and the maximum bandwidth of a multimode waveguide of length L is

$$\Delta\nu < \frac{2nc}{A^2 L}. \tag{12.20}$$

For a 1-km multimode fiber with numerical aperture $A = 0.2$ and $n = 1.5$, the bandwidth is limited to $\Delta\nu \ll 22.5$ MHz.

In conclusion, single-mode fibers are essential for long-distance, high-bandwidth optical communication because they eliminate modal dispersion and can be engineered to minimize material dispersion. In contrast, multimode fibers are limited by intermodal delays, which restrict their usable bandwidth and make them more suitable for short-distance applications such as local area networks and imaging systems.

12.6 Applications of Optical Waveguides

Optical waveguides enable controlled propagation of light over long distances. Their applications include telecommunications, sensing, imaging, and lasers.

In optical communications, waveguides transmit data with low loss and high bandwidth over thousands of kilometers. They support wavelength-division multiplexing, when several channels are compressed in a single fiber, which significantly increases channel capacity.

In sensing applications, waveguides are used to detect changes in physical properties in the ambient near the guiding surface in chemical and biological sensors, with applications in medical diagnostics and environmental monitoring.

Integrated optics relies on planar waveguides fabricated on substrates such as silicon or lithium niobate to produce integrated modulators, splitters, interferometers, and resonators with applications in quantum technologies, signal processing, and optical computing.

Fiber lasers are used in telecommunications, precision manufacturing, medicine, and defense, with power levels ranging from from milliwatts to tens of kilowatts.

In imaging systems, waveguides transmit images in image intensifiers and endoscopes.

Photonic crystal fibers offer novel light-guiding mechanisms for advanced laser systems.

12.7 Conclusion

Optical waveguides confine and transport light through total internal reflection. The numerical aperture defines the maximum acceptance angle for guided modes and, together with the waveguide dimensions, specifies the number of supported modes.

In step-index cylindrical fibers, the V-parameter defines whether the fiber operates in single- or multimode regime. Single-mode operation occurs at $V \leq 2.405$. The number of modes supported by a multimode waveguide is defined as $V^2/2$.

Efficient coupling of free space sources to waveguides requires matching the numerical aperture and spatial profile of the source to the one of the guided mode. The overlap integral provides a measure of coupling efficiency.

Bandwidth of the guided wave is limited both by material dispersion, and by intermodal dispersion in multimode fibers. These effects restrict multimode fibers to short links. Single-mode fibers, with dispersion-engineered glass compositions, support transmission over hundreds of kilometers at gigahertz bandwidths.

12.8 Problems

12.1 Assume the fundamental mode of a fiber is described by Gaussian distribution

$$M(\rho) = \exp(-\rho^2),$$

where ρ is the radial coordinate normalized to the waist of the waveguide mode. The coupling field is described by

$$E(\rho) = \exp(-\frac{\rho^2}{w^2}),$$

where w is the width of the gaussian beam to be coupled. Estimate the coupling efficiency as a function of w.

Solution (12.1) Using 12.14:

$$T = \frac{[\int_0^\infty \exp(-\rho^2)\exp(-\frac{\rho^2}{w^2})\rho d\rho]^2}{\int_0^\infty \exp(-2\rho^2)\rho d\rho \int_0^\infty \exp(-\frac{2\rho^2}{w^2})\rho d\rho} = \frac{4}{w^2+1} - \frac{4}{(w^2+1)^2}.$$

The graph of $T(w)$ is shown in Fig. 12.3.

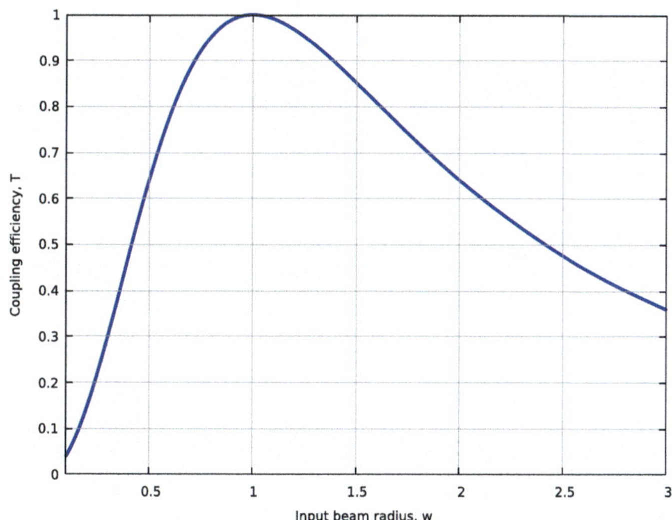

Fig. 12.3 To problem 12.1: coupling efficiency as a function of beam waist w normalized to the radius of the waveguide mode

Chapter 13
Adaptive Optics

Abstract Adaptive optics is introduced as a technology for dynamic correction of rapidly changing aberrations in optical systems. Its three major components—the wavefront corrector, the wavefront sensor, and the control system are defined, and the function of each is explained. The requirements for a wavefront corrector used in compensating atmospheric turbulence are established based on the Kolmogorov turbulence model. The shape of a deformable mirror's influence functions is analyzed through the principles of solid-state mechanics. Major actuator types and various deformable mirror designs are described. Liquid crystal wavefront correctors are presented as an alternative technology for high-resolution wavefront correction. Finally, Alvarez optical components are introduced as a solution specifically designed to provide single-mode correction of predetermined aberrations.

13.1 Architecture of an AO System

Aberrations in optical systems are the primary source of image degradation. Efforts to compensate for static aberrations led to significant advances in optical materials, fabrication methods, and optical design theory, which had largely matured by the mid-20th century. However, the compensation of dynamic aberrations remained an unsolved problem at that time.

Atmospheric turbulence is the dominant factor limiting the resolution of optical systems operating through the atmosphere. Babcock [1] in the USA and Linnik [2] in the Soviet Union independently proposed dynamically controlled optical elements to compensate for rapidly varying aberrations caused by turbulence.

The concepts suggested by Babcock and Linnik already included the fundamental building blocks of an adaptive optics (AO) system—namely, the corrector, the sensor, and the controller.

In a broader sense, adaptive optics (AO) is a technology for real-time dynamic correction of distortions and aberrations in optical systems. In many cases AO allows for relaxed precision requirements in conventional optics, but at the same time it imposes stringent demands on the specialized components of the AO loop, including

G. Vdovin, *Elementary Technical Optics*, UNITEXT for Physics,
https://doi.org/10.1007/978-3-032-08626-6_13

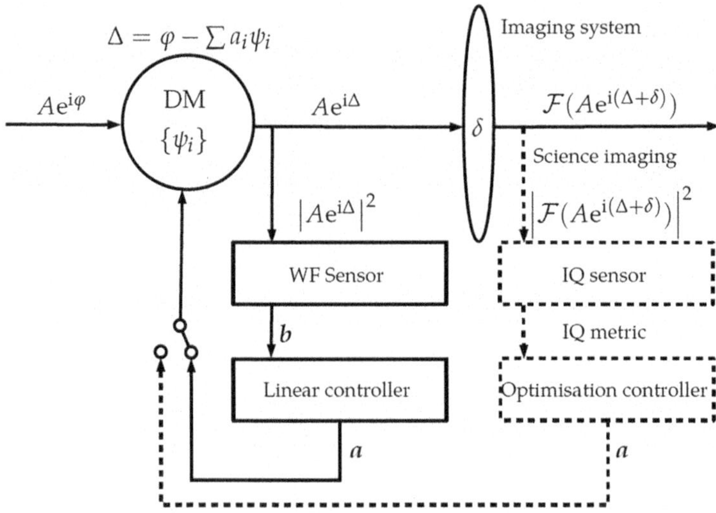

Fig. 13.1 General schematic of an AO system

wavefront correctors, wavefront sensors, and control algorithms. Because of this complexity, AO development has historically advanced in three parallel directions: sensing, correction, and control.

Figure 13.1 illustrates the flow of physical signals in a typical AO system. A distorted optical wave with amplitude A and phase $\varphi(\bar{\rho}, t)$ is modified by a deformable mirror (DM). Unlike static optical elements, which influence both amplitude and phase, the DM introduces controlled phase corrections represented as a linear combination of its influence functions $\psi_i,\ i = 1, \ldots, N$. After reflection, the corrected residual phase error is

$$\Delta = \varphi - \sum a_i \psi_i .$$

In an AO system, the central goal is to sense the wavefront and compute the control vector $\bar{a}$ that minimizes Δ. An ideal optical system producing a perfect image corresponds to the condition $\Delta = 0$.

The deformable mirror must provide precise phase corrections without introducing absorption or scattering. Its initial surface flatness must be extremely high to avoid unwanted distortion, and the mirror must be capable of a broad correction spectrum to minimize residual error Δ across a wide range of turbulence statistics. Additionally, the correction speed must be high to reduce temporal errors. The combined requirements of high reflectivity, broad correction range, temporal and thermal stability make wavefront correction a demanding and continually evolving technology.

Direct sensing of the optical phase is impossible because of the extremely high frequency of light. Therefore, AO systems rely on indirect methods that reconstruct phase information from intensity measurements $|Ae^{i\Delta}|^2$. Recovering phase from intensity is an ill-posed problem, requiring additional information or specific arrangements. This challenge complicates AO control, which must either employ methods that linearize the relation between intensity and phase, or solve a non-linear optimization problem that maximizes a performance metric of $|Ae^{i\Delta}|^2$ by adjusting Δ.

Recent developments in astronomy, microscopy, and biomedical imaging have created demands for AO systems with tens of thousands of degrees of freedom. Direct real-time solutions of the corresponding large-scale linear problems are computationally intractable even with modern hardware. Furthermore, classical linear AO schemes minimize phase error in the wavefront sensor plane, while image quality is also affected by aberrations introduced outside the sensing loop. As a result, conventional AO cannot correct so-called non-common path errors δ.

In many applications, phase can be indirectly optimized by maximizing an image-plane metric, such as image sharpness or beam brightness. This approach defines the class of sensorless AO systems. In the simplest case, the intensity of a coherent optical system can be written as the squared modulus of the Fourier transform of the pupil field:

$$I \sim |\mathcal{F}[A\exp(i(\Delta + \delta))]|^2 .$$

Here the final image naturally incorporates the influence of non-common path error δ, often yielding superior correction quality compared to linear pupil-sensing systems. However, optimization in sensorless AO is also ill-posed and requires iterative control methods. We use the term "sensorless" to describe systems that operate without a dedicated pupil-plane wavefront sensor, relying instead on direct feedback from an image quality metric. Convergence requires multiple correction steps, typically more than the number of degrees of freedom.

The architecture of AO systems must also consider structural optimization. For example, defocus aberration can in principle be corrected by a deformable mirror with many actuators, but theory shows that a uniform pressure applied to a circular membrane produces the exact correction shape. Thus, in theory, a single-actuator membrane mirror could correct defocus. In practice, however, such a design would be unable to compensate additional aberrations from misalignments or imperfections. Therefore, real systems employ multiple actuators: the global defocus signal is distributed as predicted by theory, while fine corrections are applied through sensing and control tailored to the specific task.

Applications of AO systems are extremely broad, spanning high-power laser systems, biomedical imaging, microscopy, surveillance, astronomy, ultrafast optics, and more. The diversity of processes controlled by AO requires a correspondingly wide range of control strategies and algorithms, far beyond the basic architecture of Fig. 13.1. These methods must incorporate accurate models of turbulence (e.g., Kolmogorov statistics), wavefront correctors (e.g., solid-state mechanics), and sensing techniques (both coherent and incoherent), in order to achieve robust and application-specific adaptive correction.

In practice, both refraction and reflection are used to fabricate correction elements. It was showsn in Sect. 2.4 that the deformation of mirror surface ΔM doubles the deformation of the reflected wavefront $\Delta W = -2\Delta M$, while the deformation of the thickness of a transmitting element ΔT produces wavefront deformation of $\Delta W = (n-1)\Delta T$. Reflective AO has much higher sensitivity, and has to satisfy to higher quality demands.

13.2 Quality of Adaptive Correction

In an incoherent imaging setup, the optical system forms an image of the object. Each object point acts as an independent point source, emitting (or reflecting) light in all directions and forming a spherical wavefront. In an ideal optical system, these divergent spherical wavefronts are converted into convergent spherical wavefronts. All rays, orthogonal to the spherical wavefront, intersect at a single point, forming a sharp image. Since an object consists of many points, the ideal optical system forms separate undistorted convergent spherical waves for each point. The greater the deviation of the wavefront from the ideal sphere, the less sharp the image will be.

The Strehl parameter St is the basic metric characterizing optical quality:

$$\mathrm{St} = e^{-\sigma^2}, \qquad \sigma \ll 1, \tag{13.1}$$

where σ is the *rms* wavefront deviation from the ideal sphere, measured in radians. St represents the axial intensity in the focus of an aberrated optical system, relative to the intensity I_0 produced by an ideal unaberrated beam in the far field:

$$I_0 = \frac{\pi^2 (D/2)^4}{\lambda^2 L^2} I_{\mathrm{a}}, \tag{13.2}$$

where D is the aperture diameter, L is the distance, and I_{a} is the intensity in the aperture.

The expression 13.1 is valid only for $\sigma \ll 1$. However, for atmospheric turbulence following Kolmogorov [3, 4] statistics, the following expression provides a reasonable approximation to the average Strehl value for $\sigma \geq 1$:

$$\mathrm{St} \approx \frac{1}{(1+\sigma^2)^{6/5}}. \tag{13.3}$$

This expression should be used with caution, because large aberrations produce random speckle intensity fields, causing large uncertainties in point-wise measured intensities, including on-axis values.

For diffraction-limited performance, the requirements can be derived from the Strehl parameter. To satisfy the condition $St > 0.8$, the wavefront error must be $\sigma < 0.5$ rad, which corresponds to ~20 nm *rms* surface error in the visible range.

To characterize atmospheric distortion strength, Fried [5] introduced the parameter r_0, later named as "Fried radius", defined as the aperture diameter at which the *rms* aberration equals 1 radian. The total aberration within an aperture of diameter D is then a function of the ratio D/r_0.

Noll [6, 7] used atmospheric turbulence statistics to calculate the *rms* aberration in terms of an infinite series of Zernike polynomials. The total wavefront error is

$$\sigma^2 = \left(\frac{D}{r_0}\right)^{5/3} \sum_{2}^{N} c_j, \tag{13.4}$$

where $c_{2,3} = 0.449$ (tip-tilt), $c_{4,5,6} = 0.0232$ (defocus and astigmatism), and $c_{7,8} = 0.00619$ (coma). The piston term c_1 does not affect image quality.

The residual aberration after correcting N terms is proportional to

$$\Delta_N = \sum_{N}^{\infty} c_j. \tag{13.5}$$

Noll obtained $\Delta_1 = 1.0299$, $\Delta_3 = 0.134$, $\Delta_{10} = 0.0401$, and in general

$$\Delta_N \approx 0.2944\, N^{-\sqrt{3}/2}, \quad N > 10.$$

From this we can calculate the necessary deformation range of a deformable mirror r (excluding tip and tilt):

$$r \sim \frac{3\sigma}{k} = \frac{3\lambda}{2\pi} \sqrt{\Delta_3 \left(\frac{D}{r_0}\right)^{5/3}}. \tag{13.6}$$

To calculate the required number of actuators, assume the number of corrected terms in 13.5 exceeds 10. To reduce the residual error below 1 rad ($\sigma < 1$), the minimum number of Zernike terms must satisfy:

$$\left(\frac{D}{r_0}\right)^{5/3} \cdot 0.2944\, N^{-\sqrt{3}/2} = 1, \tag{13.7}$$

with solution $N > 0.25(D/r_0)^{1.92}$. Since real deformable mirrors are less efficient than ideal Zernike correctors, a safe engineering estimate is:

$$N \geq \left(\frac{D}{r_0}\right)^2. \tag{13.8}$$

With the actuator number defined, the number of sub-apertures in a Hartmann–Shack sensor used in the AO loop should be at least two to three times larger than N, for stable control.

Finally, the response time for tip–tilt correction τ_{tt} can be estimated from the aperture size and wind speed v:

$$\tau_{tt} \approx \frac{D}{v}, \tag{13.9}$$

while the response time τ for higher-order correction is:

$$\tau \approx \frac{r_0}{v}. \tag{13.10}$$

The optical system should first be designed for maximum quality without adaptive correction. If dynamic correction is still required, further analysis must determine whether adaptive correction will improve system performance.

Adaptive correction requires precise coordination between wavefront sensing, deformable mirror actuation, and control algorithms. The achievable image quality is limited by the residual wavefront error, actuator density, sensor resolution, and correction speed. For diffraction-limited performance, residual errors must be kept below $\lambda/10$, actuator numbers must scale at least as $(D/r_0)^2$, and the response time must match or exceed the turbulence dynamics. Thus, adaptive optics design is always a trade-off between optical performance, mechanical feasibility, and control capacity.

13.3 Influence Functions of a Deformable Mirror

In adaptive optics, an aberration a is corrected by deforming a reflective or refractive surface with N mechanical actuators. Each actuator produces a deformation described by its influence function f_i, where $i = 1 \ldots N$. The wavefront correction problem reduces to finding a combination of actuator control weights v_i that minimizes the error norm for a given set of f_i and aberration a:

$$v = \operatorname{argmin} \left| \sum_1^N v_i f_i(x, y) - a(x, y) \right|. \tag{13.11}$$

A detailed description of methods to solve this equation, and thereby generate control signals for the deformable mirror, is given in Sect. 8.7.1. From (13.11) it follows that, for each correction step, the aberration function a is decomposed into a series over the set of influence functions. The correction quality depends on the error of this decomposition. Since N is finite, it is always possible to find infinitely many functions a that are orthogonal to all f_i: $(a_i \cdot f_i)_{i=1\ldots N} = 0$. Such aberrations cannot be corrected by the deformable mirror. Fortunately, these uncorrectable aberrations usually have very low statistical weight. Thus, the set of influence functions f_i, $i = 1 \ldots N$ should provide a good statistical match to the most probable aberrations

$a_{i=1...N}$, minimizing residual error over the ensemble of expected realizations of a. Consequently, the statistics of expected aberrations form an important prerequisite for designing an effective wavefront corrector.

Karhunen-Loeve (KL) decomposition [8] (also called principal value decomposition) provides the most efficient set of functions to represent a stochastic aberration with a limited number of terms. It defines basis functions $f_i,\ \ i = 1 \ldots N$ such that the correlation matrix of decomposition coefficients with zero mean is diagonal:

$$\sum (v_i - \overline{v_i})(v_j - \overline{v_j}) = 0, \ \ i \neq j. \tag{13.12}$$

Here $v_i,\ \ v_j$ are coefficients from the solution of 13.11, and $\overline{v_i},\ \ \overline{v_j}$ are their mean values obtained by averaging over a statistically significant set of wavefront realizations. KL bases provide the fastest convergence for predefined aberration statistics. Of all deformable mirrors with N influence functions, the one with KL response achieves the best performance – but only for the given statistics of wavefront aberrations. Since the KL basis depends on the underlying statistics, the set of KL functions cannot be defined universally.

Low-order Zernike polynomials are close to Karhunen-Loeve functions derived for Kolmogorov turbulence [4]. This makes the Zernike system quasi-optimal for describing wavefront correctors, as their performance is close to that of KL-based correctors in the case of Kolmogorov turbulence, particularly when the number of terms is limited [9].

In practice, however, the influence functions of deformable mirrors or refractive correctors are determined by the geometry and mechanics of material deformation. They can differ significantly from KL functions or Zernike polynomials. The specific form of influence functions is therefore a key factor defining correction quality in adaptive optics.

In incoherent adaptive optical systems, where light has a broad spectrum and the image is an extended scene, performance is defined mainly by the ability of the deformable mirror to fit the aberration. In coherent systems, where diffraction and interference effects are critical, stricter requirements apply: the mirror surface must be free of defects to avoid scattering and speckle; the influence functions must match expected aberrations to prevent hot and cold spots; and segmented mirrors are rarely used for lasers because scattering, interference, and energy loss at segment gaps are intolerable.

As shown in Sect. 2.4, to correct the aberration, the mirror surface must reproduce the aberration with half its amplitude and opposite sign. For rapidly varying aberrations, the mirror must be dynamically deformable. Since surface deformation is caused by mechanical forces, several concepts of deformable mirrors (DMs), based on different mechanical principles, have been developed:

- Membrane mirrors, formed by a stretched reflective membrane deformed by external forces [10].
- Bimorph mirrors, formed by bonded deformable plates that bend under induced strains [11].

- Continuous facesheet mirrors, consisting of a thin reflective plate controlled by an array of push-pull actuators [12].

The shape ϕ of a membrane mirror with contour S is described by the Poisson equation:

$$\Delta\phi(x, y) = p(x, y),\quad F|_S = 0, \tag{13.13}$$

where p is the electrostatic pressure applied by the actuators. The same equation describes bimorph mirrors, with p representing bending moments. The boundary conditions for bimorph mirrors may vary depending on design.

The shape of a continuous facesheet mirror with discrete actuators, in the thin-plate approximation $h < 10d$ (where h is thickness and d the plate diameter), is described by the biharmonic equation [15]:

$$\Delta\Delta\phi(x, y) = p(x, y), \tag{13.14}$$

where $p(x, y)$ is the applied actuator force distribution, subject to static equilibrium, and $\phi(x, y)$ satisfies additional boundary conditions. Controls p corresponding to a given phase error $\phi(x, y)$ can be found by applying $\Delta\Delta$ to ϕ. Substituting $\phi = R_n^m$, where R_n^m is a Zernike polynomial with radial order n and azimuthal order m, yields

$$\Delta\phi(x, y) = 0 \tag{13.15}$$

for all Zernike polynomials with $n = |m|$ (piston, tip, tilt, astigmatism, trefoil, etc.), and

$$\Delta\Delta\phi(x, y) = 0 \tag{13.16}$$

for all Zernike polynomials with $n = |m|$ and $n - 2 = |m|$ (piston, tip, tilt, defocus, astigmatism, trefoil, coma, and some higher-order terms).

Equations 13.13 and 13.14 describe DM shapes in terms of applied forces within the correction aperture. However, some modes depend only on boundary conditions and cannot be expressed as exact solutions of these equations.

In practice, correcting aberrations that satisfy (13.15) for membrane/bimorph mirrors, or (13.16) for thin-plate mirrors, using only actuators inside the aperture, results in high-order "bumpy" approximations. High-quality correction requires additional actuators outside the aperture, to enforce the necessary boundary conditions. The number of external actuators must be large enough to generate sufficient independent modes. For example, out of the first 15 Zernike polynomials, 14 satisfy 13.16. Thus, a continuous facesheet mirror requires at least 14 actuators outside its aperture to correct these low-order terms. For bimorph and membrane mirrors, only 9 out of 15 satisfy 13.15, so at least 10 external actuators are needed.

From basic considerations, a circular continuous facesheet mirror requires at least two rings of external actuators to define both ϕ and $\partial\phi/\partial r$ at the boundary. Bimorph and membrane mirrors need at least one ring, since only ϕ must be specified.

Equations 13.13 and 13.14 are only approximations; real mirror behavior is more complex. In general, elastic deformations of solids are governed by second- or higher-order operators, and boundary conditions are always critical. Nonetheless, the core conclusion holds: some low-order modes can only be formed by defining boundary conditions, which requires actuators placed outside the correction aperture.

Practically, this means that any mechanical deformable mirror must allocate a significant number of actuators outside the active aperture. For instance, a continuous facesheet mirror designed to correct all third-order aberrations (except spherical) must place all actuators outside the correction area.

These rules assume actuators are linear and have unlimited force. In reality, actuator clipping and saturation impose constraints, so practical DM designs deviate from these principles to balance optical performance with engineering feasibility.

Finally, since the response of a continuous facesheet DM is inherently a fourth-order (or higher) function, attempts to correct beam focus with such mirrors inevitably produce print-through effects – visible in coherent light as hot or dark spots in the actuator projection on the beam cross section. This limitation is fundamental to the continuous facesheet design.

The influence functions of a deformable mirror determine the quality and efficiency of wavefront correction. While theoretical bases such as Karhunen–Loeve functions or Zernike polynomials provide optimal or quasi-optimal descriptions for given statistics, real mirrors are constrained by mechanics, actuator placement, and boundary conditions. Therefore, practical designs must balance mathematical optimality with engineering feasibility, ensuring sufficient actuator density both inside and outside the correction aperture to achieve stable, high-fidelity correction without introducing print-through artifacts.

13.4 Actuators for Deformable Mirrors

Actuators for adaptive optics must provide a deformation range of several wavelengths (up to tens of wavelengths in some applications), fast response, high linearity, low hysteresis, excellent repeatability, long-term stability, and reproducible response.

The required deformation range depends on the application. For atmospheric turbulence correction, a range of a few microns is sufficient, while correction of strongly aberrated human eyes requires tens of microns [13]. The final precision of wavefront correction is typically on the order of $\lambda/10$, and for demanding applications such as optical lithography or interferometry, as fine as $\lambda/100$.

Although no single actuation technology meets all requirements, several approaches are commonly used:

- *Electrostatic actuators*, employed in membrane and micromachined deformable mirrors, provide a response proportional to the square of the applied voltage. Their actuation does not depend on the sign of the control signal and they feature negligible hysteresis.
- *Piezoelectric actuators*, based on the piezoelectric effect where an electric field alters the polarization of the material to produce mechanical strain. Their response is linear with the control voltage, but they exhibit significant hysteresis and creep, making the overall response nonlinear.
- *Electromagnetic actuators*, available in several designs, such as solenoid or voice-coil actuators. Their response characteristics depend strongly on the specific actuator configuration.
- *Thermal actuators*, realized in multiple forms. Even ambient temperature changes can deform optical elements, but more controlled devices exploit localized heating. Thermal actuators have been successfully implemented in deformable mirrors [14]. Since their response time scales with the square of the actuator size, micromachined thermal actuators with very small dimensions can reach response frequencies of tens to hundreds of Hz.

13.5 Overview of DM Technologies

Although the concept of deformable mirrors has been known for decades, practical real-time wavefront correction systems used to be complex and costly. Historically, adaptive optics and deformable mirrors were primarily developed for military applications such as satellite tracking and imaging through turbulence, and for astronomy. With the advent of microtechnologies, inexpensive deformable mirrors became available [10].

Today the market offers a wide range of relatively mature and affordable DMs, based on different physical principles, with actuator counts reaching into the thousands.

13.5.1 Continuous Faceplate DMs with Push-Pull Actuators

The concept of a piezoelectric DM with push-pull actuators, first introduced more than 40 years ago [12], is illustrated in Fig. 13.2. The design consists of a stable, rigid baseplate with a grid of piezoelectric actuators attached, supporting a thin deformable faceplate on top. Each actuator locally deforms the mirror surface.

Piezoelectric DMs employ relatively expensive actuators. They require high driving voltages and peak currents, resulting in significant power dissipation in the driver electronics, with additional heat losses in the DM itself due to actuator hysteresis. Their high cost and limited reliability hinder scalability, although custom devices for extremely large telescopes have been built with thousands of actuators.

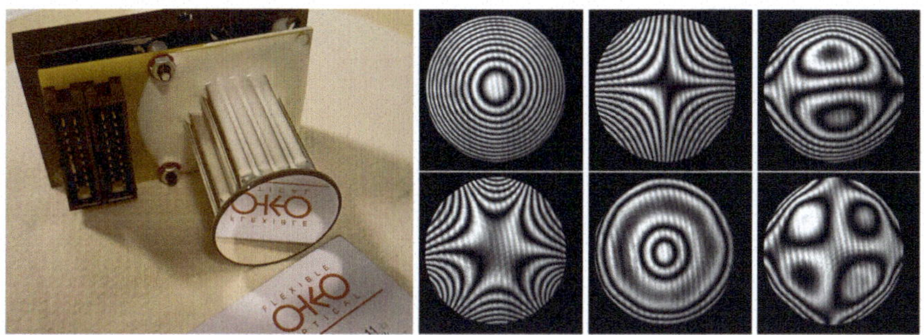

Fig. 13.2 30-mm 37-channel piezoelectric deformable mirror (package removed) and interferograms of low-order Zernike modes produced with it

Continuous faceplate mirrors are highly versatile correctors suitable for many applications. Their relatively thick faceplates can be coated not only with metal films but also with multilayer high-reflectivity dielectric stacks, making them well suited for wavefront correction and beam shaping in high-power laser systems.

Flexibility in the geometry of actuator array design makes them attractive for custom devices where only specific aberrations need to be corrected, but with high precision.

However, these mirrors also present drawbacks that require careful design compromises:

- Actuator hysteresis varies by technology: nearly zero for thermal actuators, $\sim 2\%$ for electrostrictive actuators, and 10–15% for piezoelectric actuators.
- Piezoelectric stacks tolerate only 2–5% of their maximum compressive stress in tension. Since actuators in DMs operate under both tensile and compressive stress, careful design is required to avoid premature failure.
- Considerable *print-through* often occurs. The larger the DM stroke, the stronger the print-through. This effect arises from local faceplate deformation near actuator attachment points and cannot be corrected by the DM itself, thus contributing to residual wavefront error.

Implementation of continuous faceplate deformable mirrors must take into account several critical factors, such as stress-free assembly of the deformable plate with the actuator stack, minimization of coating stress, achieving a high-quality initial figure, and incorporation of an efficient cooling system for high-power operation. This list, while already extensive, is still incomplete, as many practical challenges remain. Nevertheless, a simple mechanical model of a continuous faceplate DM is very useful for preliminary estimation of the amplitude and shape of the DM influence functions, as well as for assessing correction performance.

A 37-channel piezoelectric deformable mirror and interferograms of Zernike polynomials generated with this type of DM are shown in Fig. 13.2.

Piezoelectric DMs with push-pull actuators are commercially produced by Xinetics (USA), CILAS (France), and Flexible Optical BV (The Netherlands) [16].

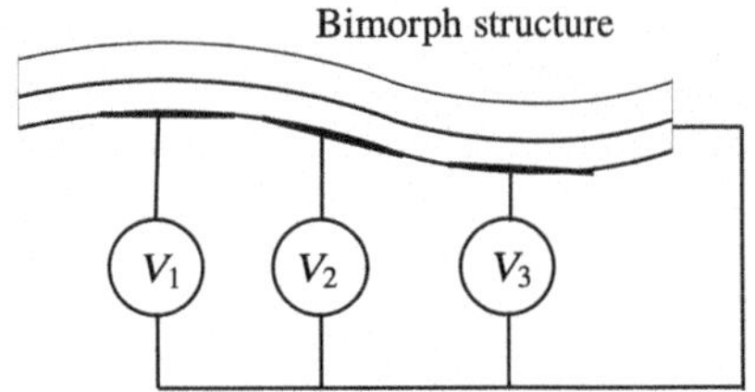

Fig. 13.3 Principle of a bimorph DM

13.5.2 Bimorph DM

The bimorph deformable mirror [11], illustrated in Fig. 13.3, consists of two or more bonded layers of different (typically piezoelectric) materials. Lateral deformation of the patterned active layer produces changes in the local curvature of the reflective surface. Bimorph DMs provide accurate reproduction of low-order Zernike modes, but only limited correction amplitude for higher-order modes. Consequently, most bimorph mirrors on the market are built with fewer than 100 actuators.

Because the maximum inter-actuator deformation in a bimorph mirror is inversely proportional to the number of degrees of freedom, the practical actuator count is usually below 100. As with all piezoelectric DMs, bimorph mirrors exhibit significant hysteresis.

Commercial bimorph DMs are manufactured by CILAS (France), NighN (Russia), and Thorlabs (USA).

13.5.3 Membrane Deformable Mirror

Membrane deformable mirrors, example of which is illustrated by Fig. 13.4, generally employ electrostatic actuators. These actuators operate by electrostatic attraction between two charged surfaces, without requiring any bulk material in the active gap. Since material properties (apart from the membrane itself) are not involved, electrostatic actuators feature negligible hysteresis and, in theory, consume no static power.

Key properties of electrostatic actuation are:

- The force is always attractive; implementation of a repulsive electrostatic actuator is technically challenging and requires use of electret materials.
- The force is proportional to the square of the electric field in the gap.
- Because the field depends on both applied voltage and the gap size—which changes during deformation—electrostatic actuators exhibit nonlinear response, even after being linearized for quadratic respone. Fortunately, the response is free of hysteresis.

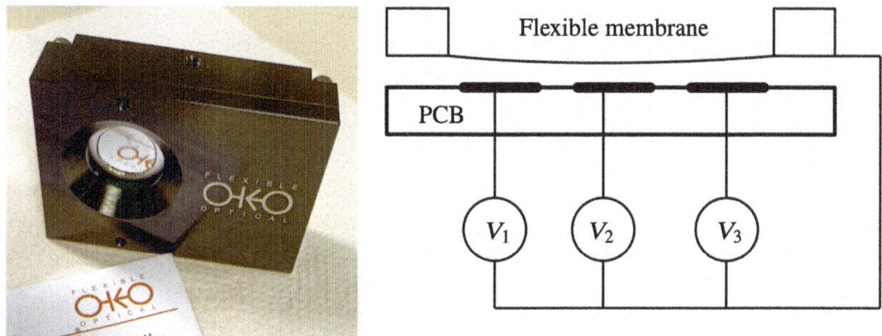

Fig. 13.4 MEMS DM with electrostatic control (OKO) and schematic

Similar to bimorph mirrors, membrane DMs provide large deformations for low-order Zernike modes and only limited correction amplitude for higher-order modes. Because maximum inter-actuator deformation decreases with increasing degrees of freedom, the number of actuators in commercial membrane mirrors is typically fewer than 100.

Membrane DMs are produced by Alpao (France), Imagine Eyes (France), Adaptica (Italy), and Flexible Optical BV (The Netherlands).

13.5.4 Micromachined DM

Micromachined deformable mirrors (MEMS DMs) operate on the same physical principles as macroscopic DMs but are partly or entirely fabricated using micromachining technology.

Boston Micromachines [17] developed a continuous faceplate MEMS DM with push-pull actuators, as shown in Fig. 13.5. These devices represent the most advanced DMs currently available, with actuator counts reaching up to 4,000. However, each actuator requires a dedicated high-voltage connection to an external driver amplifier. The complexity of external wiring fundamentally limits scalability beyond tens of thousands of actuators.

Micromachined DMs are produced by Boston Micromachines (USA) and Flexible Optical BV (The Netherlands).

13.6 Liquid Crystal Wavefront Correctors

Liquid crystals (LC) [18] combine properties of both liquids and crystals: they flow like liquids but at certain conditions maintain ordered crystalline structure. For certain classes of LC materials, the refractive index strongly depends on the polarization and the propagation direction of light with respect to the molecular orientation. Since LC molecular alignment can be controlled by boundary conditions or by external electric fields, LC devices can be used to manipulate both the intensity and the phase of light.

The most common application of liquid crystals is intensity control, as in LC displays, which have achieved a very high level of technological maturity.

LC phase modulators are similar in construction to intensity modulators, but impose stricter requirements on layer thickness, molecular orientation, and control electronics. Nematic liquid crystals, consisting of long molecules, can be aligned either mechanically (using nano-structured alignment layers) or electrically. A typical LC modulator cross-section with orientation layers is shown in Fig. 13.6.

The cell consists of two transparent glass substrates with indium–tin–oxide (ITO) electrodes that define the actuator structure. These electrodes are coated with orientation layers to control the initial LC alignment. With no voltage applied, molecules align uniformly along the surface. When a voltage is applied, molecules reorient along the electric field. This changes the effective refractive index experienced by polarized light, shifting it between the extraordinary and ordinary indices. The birefringence can reach values as high as $\Delta n = n_e - n_o \approx 0.2$. The induced birefringence depends on voltage in a predictable, although nonlinear, manner.

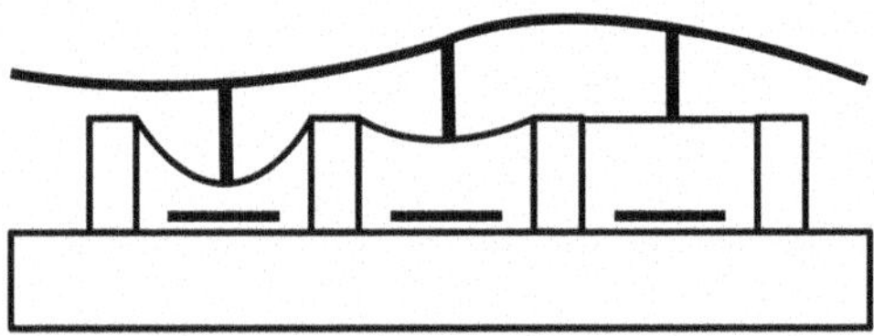

Fig. 13.5 Structure of MEMS DM from Boston Micromachines. The intermediate membrane is sectioned to form electrostatic actuators, whose motion is transferred to the continuous flexible faceplate via micromachined support columns

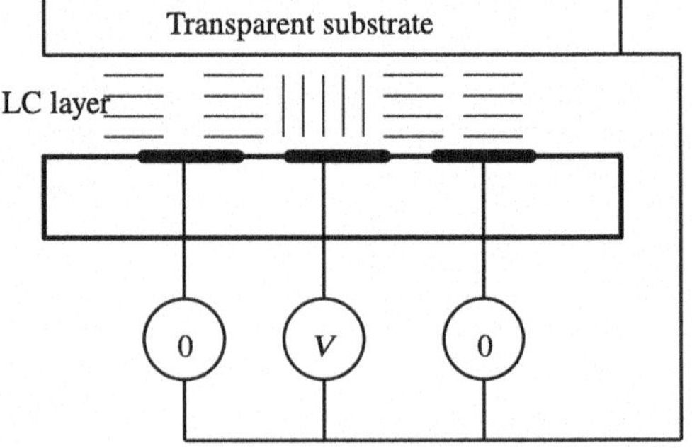

Fig. 13.6 Operation of an LC phase modulator. At zero voltage, LC molecules align along the gap; when voltage is applied, the molecules reorient, changing the local refractive index

The phase modulation depth achievable by a thin LC layer is given by:

$$\Delta\varphi = \frac{2\pi\,\Delta n t}{\lambda}, \tag{13.17}$$

where t is the LC layer thickness.

For full 2π phase modulation, the LC layer thickness must satisfy $t > \lambda/\Delta n$, which typically corresponds to 2–50 μm. However, thicker layers slow down switching because the response time τ scales as $\tau \sim t^2$. Molecular order is also less stable in thicker layers, so modern LC devices are designed with the thinnest possible layer still capable of 2π modulation at the target wavelength. Devices optimized for longer wavelengths must use thicker LC layers and thus operate more slowly.

LC modulators can be fabricated with millions of pixels. Such high spatial resolution allows fine wavefront control with some parasitic scattering but with relatively high diffraction efficiency.

An alternative approach is to apply smooth electric field distributions across the LC, producing continuous refractive index variations with reduced scattering. However, this requires thicker layers and results in slower response. LC lenses of this type are described in [19].

In summary, LC modulators offer very high resolution, low voltage operation, and minimal power consumption. Their disadvantages are slow response, polarization dependence, and the 2π phase-wrapping problem, which restricts their use mainly to monochromatic applications.

13.7 Generalized Alvarez Correctors

Alvarez [20] proposed a variable-focus optical system consisting of two identical phase plates laterally shifted relative to each other, as shown in Fig. 13.7. This principle has been widely adopted in technical and ophthalmic optics. Later extensions include generation of variable coma and spherical aberrations [21].

Consider two aspherical plates: the first introduces aberration $f(x, y)$, and the second introduces $-f(x, y)$. When aligned, the plates cancel each other exactly: $f(x, y) - f(x, y) = 0$. If shifted laterally by δ along the x-axis, the combined aberration becomes:

$$f(x + \delta/2, y) - f(x - \delta/2, y).$$

To generate a desired aberration $F(x, y)$, the following condition must be satisfied:

$$F(x, y) = f(x + \tfrac{\delta}{2}, y) - f(x - \tfrac{\delta}{2}, y) \approx \delta \frac{\partial f(x, y)}{\partial x}. \tag{13.18}$$

Thus, the required phase plate function is:

$$f(x, y) = \frac{1}{\delta} \int F(x, y) dx. \tag{13.19}$$

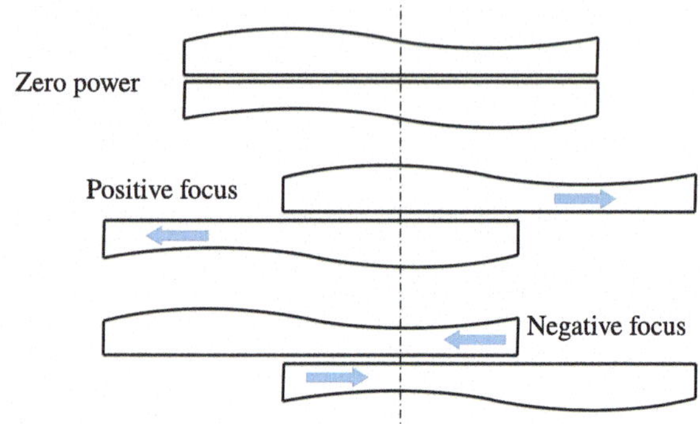

Fig. 13.7 Principle of an Alvarez variable-focus lens

Equation 13.19 directly leads to the design of the Alvarez variable-focus lens, where defocus is achieved by shifting the two plates relative to each other.

Example: Design an Alvarez lens made of glass with refractive index $n = 1.5$. The lens should provide zero focusing power at zero shift, and a focusing power of 1 diopter at a relative shift of 1 mm.

The corresponding wavefront with a radius of $R = 1$ m is

$$F(x, y) = \frac{x^2 + y^2}{2R} = 0.5(x^2 + y^2),$$

where x and y are in meters. From Eq. 13.19, the required phase plate function is:

$$f(x, y) = \int \frac{F(x, y)}{\delta} dx = \int \frac{0.5(x^2 + y^2)}{10^{-3}} dx = 500 \left(\frac{x^3}{3} + xy^2\right).$$

The optical surface is obtained by dividing by $(n - 1)$:

$$S(x, y) = \frac{f(x, y)}{n - 1}.$$

For $n = 1.5$, we obtain

$$S(x, y) = 1000 \left(\frac{x^3}{3} + xy^2\right), \qquad x, y \text{ in meters.}$$

This surface equation describes the optical profile required for each Alvarez plate.

13.8 Conclusion

Adaptive optical systems are used for dynamic compensation of optical aberrations through wavefront sensing and correction. The deformable mirror is the most frequently used wavefront corrector. The response functions of the deformable mirror determine the range of correctable aberration. The number of actuators must scale with the square of D/r_0 to achieve diffraction-limited performance. Residual error of correction strongly depends on the actuator geometry, boundary conditions, and mechanical constraints. Making optimization critical for the design of deformable mirrors.

Wavefront sensors, such as Hartmann–Shack arrays, retrieve phase information indirectly by sampling the field in the pupil, whereas sensorless approaches optimize directly in the image plane using performance metrics such as sharpness or brightness. The latter naturally incorporate all system errors, but require iterative algorithms and longer convergence times. For both approaches, computational demands increase rapidly with actuator count.

Piezoelectric stack actuators provide high stroke in a deformable mirror, but suffer from hysteresis; electrostatic actuators are compact and hysteresis-free but limited in force; thermal actuators offer simplicity but feature slow response; MEMS implementations achieve thousands of actuators with compact integration but face scalability constraints.

Liquid crystal modulators deliver very high spatial resolution at low voltage, limited by response time and polarization dependence. Alvarez elements provide efficient single-mode correction of predefined aberrations.

In practice, adaptive optics integrates optical mechanical and computer control technologies. The choice of corrective architecture is dictated by the nature of aberration, aperture size, and quality requirements. Nowadays adaptive optics is widely used in astronomy, laser physics, medicine and high-resolution imaging.

13.9 Problems

13.1 Formulate the requirements for the wavefront corrector (number of actuators, maximum stroke, response time) to be used with a telescope with aperture of 1 m, assuming wavelength of $\lambda = 10^{-6}$ m, $r_0 = 0.05$ m and the maximum wind velocity of 10 m/s.

Solution (13.1) Using 13.6, 13.8 and 13.9 we obtain:

- Range of correction $r > 2.1 \cdot 10^{-6}$ m,
- Minimum number of Zernike terms: $N \approx 80$,
- Number of actuators: $N = 400$,
- Response time for tip-tilt: $\tau_{tt} \approx 0.1$ s,
- Response time at highest frequency: $\tau \approx 0.005$ s.

13.2 For a specific liquid crystal material, the difference between the extraordinary and ordinary refraction $\delta_n = n_e - n_o = 0.3$. The total thickness of the layer is $t = 20$ micron, the full aperture is $D = 5$ mm. Calculate the mimimum focal length F of a lens that can be obtained by smooth parabolic modulation of the LC orientation in polarized light. How the value of F will change if we change the aperture to 2.5 mm?

Solution (13.2) The maximum optical path difference equals $\Delta = \delta_n \cdot t = 6 \cdot 10^{-6}$ m. The focal length is given by

$$F = \frac{(D/2)^2}{2\Delta} \approx 0.52\text{m}.$$

If we the aperture is reduced by two times, the focal length will become four times shorter.

13.3 Design Alvarez corrector, producing 1 μm of coma in the aperture with a diameter of $D = 2$ cm, when the components are shifted by $\delta = 1$ mm with respect to each other.

References

1. H.W. Babcock, The possibility of compensating astronomical seeing. Publ. Astron. Soc. Pac. **65**(386), 229–236 (1953)
2. V.P. Linnick, On the principal possibility of reducing the influence of the atmosphere on the star image (in Russian). Opt. Spectrosc. **3**(4), 228–232 (1957)
3. A.N. Kolmogorov, The local structure of turbulence in incompressible viscous fluid for very large Reynolds' numbers. Dokl. Akad. Nauk SSSR **30**, 301–305 (1941)
4. A.N. Kolmogorov, A refinement of previous hypotheses concerning the local structure of turbulence in a viscous incompressible fluid at high Reynolds number. J. Fluid Mech. **13**(01), 82–85 (1962)
5. D.L. Fried, Optical resolution through a randomly inhomogeneous medium for very long and very short exposures. J. Opt. Soc. Am. **56**(10), 1372–1379 (1966)
6. R.J. Noll, Zernike polynomials and atmospheric turbulence. J. Opt. Soc. Am. **66**(3), 207–211 (1976). https://doi.org/10.1364/JOSA.66.000207
7. L.C. Andrews, *Field Guide to Atmospheric Optics* (SPIE Press, 2004)
8. A. Hyvarinen, J. Karhunen, E. Oja, *Independent Component Analysis* (Wiley, 2001)
9. G. Dai, Modal wave-front reconstruction with Zernike polynomials and Karhunen-Loève functions. J. Opt. Soc. Am. A **13**(6), 1218–1225 (1996). https://doi.org/10.1364/JOSAA.13.001218
10. G. Vdovin, P. Sarro, Flexible mirror micromachined in silicon. Appl. Opt. **34**(16), 2968–2972 (1995)
11. E. Steinhaus, S.G. Lipson, Bimorph piezoelectric flexible mirror. J. Opt. Soc. Am. A **69**, 478–481 (1979)
12. R.H. Freeman, J.E. Pearson, Deformable mirrors for all seasons and reasons. Appl. Opt. **21**(4), 580–588 (1982). https://doi.org/10.1364/AO.21.000580
13. L. Thibos, R.A. Applegate, J.T. Schwiegerling, R. Webb, Standards for reporting the optical aberrations of eyes, in *Vision Science and its Applications* (SuC1. Optical Society of America, 2000)
14. G. Vdovin, M. Loktev, Deformable mirror with thermal actuators. Opt. Lett. **27**(9), 677–679 (2002). https://doi.org/10.1364/OL.27.000677

15. S. Timoshenko, S. Woinowsky-Krieger, *Theory of Plates and Shells* (McGraw-Hill, 1959)
16. Flexible Optical BV. http://okotech.com
17. T.G. Bifano, J. Perreault, R.K. Mali, M.N. Horenstein, Microelectromechanical deformable mirrors. IEEE J. Sel. Top. Quantum Electron. **5**(1), 83–89 (1999). https://doi.org/10.1109/2944.748109
18. P. Yeh, C. Gu, *Optics of Liquid Crystal Displays* (Wiley, 2010)
19. A.F. Naumov, M.Y. Loktev, I.R. Guralnik, G. Vdovin, Liquid-crystal adaptive lenses with modal control. Opt. Lett. **23**(13), 992–994 (1998). https://doi.org/10.1364/OL.23.000992
20. L.W. Alvarez, Two-element variable-power spherical lens. *US Patent 3305294* (1967)
21. N. López-Gil, H.C. Howland, B. Howland, N. Charman, R. Applegate, Generation of third-order spherical and coma aberrations by use of radially symmetrical fourth-order lenses. J. Opt. Soc. Am. A **15**(9), 2563–2571 (1998). https://doi.org/10.1364/JOSAA.15.002563

Chapter 14
Simulation of the Propagation of Light

Abstract This chapter presents the fundamental computational methods used to construct numerical models of coherent optical systems. Such systems include lasers, Shack–Hartmann wavefront sensors, interferometers, and general systems using coherent optics. The core techniques rely on calculating the diffraction integral numerically using Fast Fourier transform. A spectral propagation method is derived by decomposing the wave field into a basis of plane waves. Propagation of light field via numerical solution of scalar parabolic equation is introduced as an alternative to integral-based methods. This approach offers the significant advantage of naturally incorporating the parameters of non-uniformly absorbing and refracting media. To achieve this, a parabolic equation is derived using the slowly varying amplitude approximation. Examples implemented in MATLAB are provided to illustrate the propagation and focusing of light waves. Finally, coordinate transformations are introduced as a tool to simplify the analysis of divergent and convergent beams, extending the applicability of the methods to a wider range of optical configurations.

14.1 Wave Equation in the Paraxial Approximation

The electromagnetic field propagates as a transverse wave. However, for most practical applications diffraction can be described in terms of a scalar field. Omitting the full vector notation greatly simplifies the analysis. A scalar field is therefore assumed in this text.

The wave equation in free space is given by:

$$\frac{\partial^2 F}{\partial x^2} + \frac{\partial^2 F}{\partial y^2} + \frac{\partial^2 F}{\partial z^2} = \frac{1}{c^2}\frac{\partial^2 F}{\partial t^2}, \tag{14.1}$$

where F is a function of x, y, z, and t. For a monochromatic wave we write:

$$F = \varphi \mathrm{e}^{-\mathrm{i}\omega t}. \tag{14.2}$$

G. Vdovin, *Elementary Technical Optics*, UNITEXT for Physics,
https://doi.org/10.1007/978-3-032-08626-6_14

Substituting (14.2) into (14.1) yields:

$$\frac{\partial^2 \varphi}{\partial x^2} + \frac{\partial^2 \varphi}{\partial y^2} + \frac{\partial^2 \varphi}{\partial z^2} = -k^2 \varphi. \tag{14.3}$$

Assuming propagation along the z axis:

$$\varphi = U \cdot e^{ikz}; \quad k = 2\pi/\lambda; \quad \lambda = c/w, \tag{14.4}$$

we obtain:

$$\frac{\partial^2 U}{\partial x^2} + \frac{\partial^2 U}{\partial y^2} + 2ik\frac{\partial U}{\partial z} = -\frac{\partial^2 U}{\partial z^2}. \tag{14.5}$$

Assuming U is a slowly varying function of z, the second derivative on the right-hand side can be neglected, leading to:

$$\frac{\partial^2 U}{\partial x^2} + \frac{\partial^2 U}{\partial y^2} + 2ik\frac{\partial U}{\partial z} = 0. \tag{14.6}$$

Equation 14.6 describes the propagation of a light wave in the paraxial approximation. It can be solved numerically using the method of finite differences. This equation also allows for incorporating material properties such as refraction and absorption.

14.2 Spatial Spectrum of the Light Field

Harmonic oscillation of a physical quantity U can be expressed as:

$$U = A\cos(wt - \varphi), \tag{14.7}$$

where t is time and φ is the initial phase of the oscillation at $t = 0$. Function U is a solution of the wave Eq. 14.1. Equation 14.7 describes oscillation at a fixed point in space. To describe a traveling wave, we add the coordinate dependence:

$$U = A\cos(wt - \mathbf{kr} - \varphi), \tag{14.8}$$

where A is the amplitude, $\mathbf{r}$ is the position vector, and $\mathbf{k}$ is the wave vector, whose magnitude remains to be defined. In a reference frame moving with the wave, the phase is constant, which implies that the term $wt - \mathbf{kr}$ equals zero. For a wave of wavelength λ traveling with velocity c in the direction of $\mathbf{r}$, $w = 2\pi c/\lambda$ and $|r| = ct$. The modulus of the wave vector is then:

$$|k| = 2\pi/\lambda. \tag{14.9}$$

For an optical field, the term $\cos(wt)$ represents very fast oscillations. This factor is usually omitted in formal descriptions, since it is not directly measurable. For example, the frequency of a HeNe laser is approximately $4.76 \cdot 10^{14}$ Hz. If required, the term wt can always be added later to the field description. With this simplification, expression (14.8) becomes:

$$U = A\cos(-\varphi - \mathbf{kr}). \tag{14.10}$$

Using the identity

$$\cos\phi = \frac{e^{i\phi} + e^{-i\phi}}{2}, \tag{14.11}$$

we obtain:

$$U = A \cdot \cos(-\varphi - \mathbf{kr}) = \frac{e^{i(\varphi+\mathbf{kr})} + e^{-i(\varphi+\mathbf{kr})}}{2}. \tag{14.12}$$

Expression 14.12 is a sum of two conjugates. By convention, in further analysis we retain only the first term:

$$U = Ae^{i(\varphi+\mathbf{kr})} = Ae^{i\varphi}e^{i(k_x x + k_y y + k_z z)}. \tag{14.13}$$

This simplification is justified because expression (14.13) satisfies the wave equation and thus can be directly used to describe propagating waves, with the amplitude defined by A and the phase by the argument of the exponential. Further simplification is possible by including φ in the complex amplitude A, i.e., $A = Ae^{i\varphi}$.

Expression (14.13) describes a plane wave propagating in the direction of the wave vector $\mathbf{k}$, with $|\mathbf{k}|^2 = k_z^2 + k_x^2 + k_y^2$, and components (k_x, k_y, k_z). Assuming the wave propagates at a small angle to the z axis, k_z can be expressed as:

$$k_z = \sqrt{k^2 - k_x^2 - k_y^2}. \tag{14.14}$$

Substituting this into (14.13), we obtain:

$$U = Ae^{iz\sqrt{k^2 - u_x^2 - u_y^2} + iu_x x + iu_y y}, \tag{14.15}$$

where $u_x = k_x$ and $u_y = k_y$.

Expression (14.15) can be used to construct the general solution of the wave equation $U(x, y, z)$ as a superposition of plane waves with amplitudes $A(u_x, u_y)$ and phases $\varphi(u_x, u_y)$. Such a solution is obtained by integrating over u_x and u_y:

$$U(x, y, z) \sim \iint A(u_x, u_y)e^{iz\sqrt{k^2 - u_x^2 - u_y^2}}e^{i(u_x x + u_y y)}du_x du_y. \tag{14.16}$$

At $z = 0$ expression (14.16) reduces to:

$$U(x, y, 0) \sim \iint A(u_x, u_y) \mathrm{e}^{\mathrm{i}(u_x x + u_y y)} \mathrm{d}u_x \mathrm{d}u_y. \tag{14.17}$$

Expression (14.17) has the form of an inverse Fourier transform, relating the angular spectrum of the wave $A(u_x, u_y)$ to the spatial field distribution $U(x, y, 0)$ in the plane $z = 0$.

Thus, the field $U(x, y, z)$ at a plane z can be found in two steps:

- First, compute the angular spectrum $A(u_x, u_y)$ as the Fourier transform of the field distribution $U(x, y, 0)$.
- Second, multiply the spectrum by the free-space transfer function $\mathrm{e}^{\mathrm{i}z\sqrt{k^2 - u_x^2 - u_y^2}}$ and apply the inverse Fourier transform to determine the field at plane z.

A more detailed description of the spectral representation of optical wave fields can be found in [1, 2].

14.3 FFT Propagation with the Spectral Method

Let us consider the wave function U in two planes: $U(x, y, 0)$ and $U(x, y, z)$. Suppose that $U(x, y, z)$ is obtained by propagating $U(x, y, 0)$ over a distance z. The Fourier transforms of these two fields (initial and propagated) are denoted as $A(u_x, u_y, 0)$ and $A(u_x, u_y, z)$, respectively. In the Fresnel approximation, the Fourier transform of the propagated wave function is related to the Fourier transform of the initial field via the free-space transfer function $H(u_x, u_y, z)$:

$$H = \frac{A(u_x, u_y, z)}{A(u_x, u_y, 0)} = \mathrm{e}^{\mathrm{i}z\sqrt{k^2 - u_x^2 - u_y^2}}, \tag{14.18}$$

where

$$A(u_x, u_y, 0) = \int_{-\infty}^{\infty}\int_{-\infty}^{\infty} U(x, y, 0)\, \mathrm{e}^{-\mathrm{i}k(u_x x + u_y y)} \mathrm{d}x \mathrm{d}y, \tag{14.19}$$

$$A(u_x, u_y, z) = \int_{-\infty}^{\infty}\int_{-\infty}^{\infty} U(x, y, z)\, \mathrm{e}^{-\mathrm{i}k(u_x x + u_y y)} \mathrm{d}x \mathrm{d}y. \tag{14.20}$$

Equations (14.18)–(14.20) provide a symmetric relation between the initial and propagated wave functions in the Fresnel approximation. Applied in the sequence (14.19) $\Rightarrow$ (14.18) $\Rightarrow$ (14.20), they yield the diffracted wave function, while in the reverse order they reconstruct the initial field from its propagated distribution. We denote these operations of forward and backward propagation by the operators L^+ and L^-, respectively.

This algorithm can be implemented numerically using the Fast Fourier Transform (FFT) on a finite rectangular grid with periodic boundary conditions. In this case, the model effectively describes beam propagation inside a rectangular waveguide with reflective walls, where reflections at the boundaries arise due to the periodicity of the FFT. To approximate free-space propagation, sufficiently wide guard bands must be included around the wave function within the grid. To reduce artifacts caused by the finite rectangular window, Gaussian amplitude apodization in the frequency domain should be applied.

To illustrate the method, consider a one-dimensional case.

The initial wave field is sampled into N points, forming an array defined over the interval $x = -D/2 \ldots D/2$. This interval must include both the field distribution and the required zero-filled guard bands. In most programming languages arrays are indexed from 1 to N. Since the FFT algorithm requires an even number of points, the central element of the array, corresponding to $x = 0$, is at $N/2 + 1$. Because the FFT places the zero frequency component at index 1, the array must be rearranged such that the zero frequency is shifted to $N/2 + 1$. This is achieved by the following rearrangement:

$$x_1, \ldots, x_{N/2} \rightarrow x_{N/2+1}, \ldots, x_N, \tag{14.21}$$

$$x_{N/2+1}, \ldots, x_N \rightarrow x_1, \ldots, x_{N/2}. \tag{14.22}$$

After the DFT is performed, the same rearrangement must be applied to the DFT image X_j.

The DFT image X_j is then multiplied by the free-space transfer function (14.18). To relate the index j to the frequency w_j, we use:

$$w_j = \frac{2\pi j}{D} = \frac{2\pi}{\lambda}\frac{\lambda}{D} j = k\frac{\lambda}{D} j, \tag{14.23}$$

where k is the wave number. Expression (14.23) has the physical interpretation of being the projection onto the x axis of the wave vector corresponding to a plane wave propagating at an angle $j\lambda/D$ relative to the z axis. After multiplication of the DFT of the initial field by the free-space transfer function (14.18), the propagated field in the plane z is obtained by applying the inverse FFT.

A comprehensive description of the numerical simulation of optical wave propagation can be found in [1].

The FFT spectral method provides an efficient and accurate tool for simulating optical wave propagation in homogeneous media and over large distances. Its computational efficiency makes it particularly suitable for modeling free-space diffraction and interference effects. However, because it assumes uniform refractive and absorbing properties, it is less suitable for complex or strongly inhomogeneous media, where finite-difference or beam propagation methods offer greater flexibility at the cost of higher computational effort.

14.4 Propagation with a Finite Difference Method

To illustrate the solution of the Eq. 14.6, we will use a two-dimensional (coordinates x and z) case in a free space, described by the equation:

$$\frac{\partial^2 U}{\partial x^2} + 2ik\frac{dU}{dz} = 0 \tag{14.24}$$

We re-write it as a system of finite difference equations, where index k corresponds to stepping in the z direction, and i in the x direction:

$$\frac{U_{i+1}^{k+1} - 2U_i^{k+1} + U_{i-1}^{k+1}}{(\Delta X)^2} + 2ik\frac{U_i^{k+1} - 2U_i^k}{\Delta z} = 0 \tag{14.25}$$

A higher-order approximated solution for propagation in a non-uniform medium with refraction index variation $\Delta n(x, z)$ can be obtained with modified Eq. 14.24 in the form [3]:

$$\frac{\partial^2 U}{\partial x^2} + 2ikn\frac{dU}{dz} - k^2\Delta n^2 U = 0. \tag{14.26}$$

This equation can be solved numerically by the finite difference method.

14.4.1 *Double Sweep*

Collecting terms in (14.25) we obtain the standard three-diagonal system of linear equations, solution of which describes the complex amplitude of the light field in the layer $Z + \Delta Z$ as a function of the field defined in the layer Z.

$$a_j U_{j-1}^{k+1} - c_j U_j^{k+1} + b_j U_{j+1}^{k+1} = -f_j \tag{14.27}$$

$$U_0 = k_1 U_1 + \mu_1; \quad U_N = k_2 U_{N-1} + \mu_2 \tag{14.28}$$

where we assume that the grid steps in x and y directions are equal: $\Delta X = \Delta Y = \Delta$). The three-diagonal system of linear equations (14.28) is solved by standard elimination (double sweep) method, described for example in [4]:

$$a_j = b_j = \frac{1}{\Delta^2} \tag{14.29}$$

$$c_j = \frac{2}{\Delta^2} - \frac{2ik}{\Delta Z} \tag{14.30}$$

$$f_j = -\frac{2ik}{\Delta Z} U_j^k, \tag{14.31}$$

and k_1, k_2, μ_1, μ_2 are defined from the boundary conditions.

$$\alpha_1 = k_1; \quad \beta_1 = \mu_1 \tag{14.32}$$

$$\alpha_{j+1} = \frac{b_j}{c_j - \alpha_j a_j}; \quad \beta_{j+1} = \frac{a_j b_j + f_j}{c_j - \alpha_j a_j}, \ j = 1 \ldots N-1. \tag{14.33}$$

Equations 14.32 define so called forward sweep. After all service coefficients α and β are found, the back sweep (14.34) is used to find the values of U:

$$U_N = \frac{k_2 \beta_N + \mu_2}{1 - k_2 \alpha_N} \tag{14.34}$$

$$U_j = \alpha_{j+1} U_{j+1} + \beta_{j+1}, \ j = N-1 \ldots, 1. \tag{14.35}$$

For a three-dimensional problem, with coordinates x, y and z, one step of propagation is divided into two sub-steps: the first sub-step applies the described procedure to all rows of the matrix defined in coordiantes x, y, the second sub-step changes the direction of elimination and the procedure is applied to all columns of the matrix.

The main advantage of this approach is the possibility to take into account uniformly diffraction, absorption (amplification) and refraction. In the first approximation, it can be achieved by multiplying the field, obtained after one step of propagation, by complex refractive index

$$A = t_j e^{ik(n_j - 1)\Delta}, \tag{14.36}$$

where n_j is the sampled index of refraction and t_j is the sampled function of amplitude transmission, for one z step.

As the scheme is absolutely stable (at least for free-space propagation), there is no stability limitation on the step size in the direction Z. Large steps, however, cause loss of precision, therefore the number of steps should be determined by trial (increase the number of steps in a probe model till the result stabilizes), especially for cases with strong variations of refraction and absorption inside the propagation path.

Zero amplitude boundary conditions are commonly used for the described system. This creates the problem of the wave reflection at the grid boundary. The influence of these reflections in many cases can be reduced by introducing an additional absorbing layer in the proximity of the boundary, with the absorption smoothly (to reduce the reflection at the absorption gradient) increasing towards the boundary.

14.4.2 Finite Differences for 2D Fields

Propagation of a two-dimensional field U in free space can be described by the differential equation:

$$\frac{\partial^2 U}{\partial x^2} + \frac{\partial^2 U}{\partial y^2} + 2\mathrm{i}k\frac{\partial U}{\partial z} = 0. \tag{14.37}$$

To solve this equation numerically, we re-write it as a system of finite difference equations:

$$\frac{U_{i+1,j}^{k+1} - 2U_{i,j}^{k+1} + U_{i-1,j}^{k+1}}{(\Delta X)^2} + \frac{U_{i,j+1}^{k} - 2U_{i,j}^{k} + U_{i,j-1}^{k}}{(\Delta Y)^2} + \\ +2\mathrm{i}k\frac{U_{i,j}^{k+1} - U_{i,j}^{k}}{\Delta z} = 0. \tag{14.38}$$

Here the finite differences along x and y are evaluated at different z layers, making the scheme implicit with respect to x and explicit with respect to y. This allows the system to be reduced to a one-dimensional sweep problem. Collecting terms, we obtain a standard three-diagonal system, whose solution yields the complex amplitude of the field at $Z + \Delta Z$ from its distribution at Z:

$$- a_i U_{i-1,j}^{k+1} + c_i U_{i,j}^{k+1} - b_i U_{i+1,j}^{k+1} = f_i, \tag{14.39}$$

where (assuming $\Delta X = \Delta Y = \Delta$):

$$a_i = b_i = -\frac{1}{\Delta^2}, \tag{14.40}$$

$$c_i = -\frac{2}{\Delta^2} + \frac{2\mathrm{i}k}{\Delta Z}, \tag{14.41}$$

$$f_i = \frac{2\mathrm{i}k}{\Delta Z}U_{i,j}^{k} - \frac{U_{i,j+1}^{k} - 2U_{i,j}^{k} + U_{i,j-1}^{k}}{\Delta^2}. \tag{14.42}$$

The three-diagonal system (14.39) can be solved efficiently using the elimination (double sweep) method [4]. This scheme is unconditionally stable (explicit with respect to j, implicit with respect to i). One propagation step is divided into two sub-steps: first the procedure is applied to all rows of U, then the sweep is performed along all columns.

14.5 Coordinate Transforms

Accurate numerical propagation of spherical or parabolic wavefronts in Cartesian coordinates requires very fine sampling. For example, a parabolic wave is described as

$$A\mathrm{e}^{\mathrm{i}(\phi\rho + k\frac{\rho^2}{2R})},$$

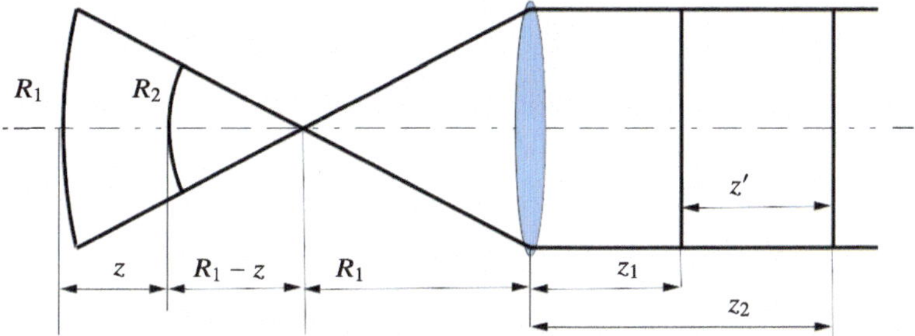

Fig. 14.1 Propagation of spherical wave with lens transform

with the large parabolic phase term

$$\varphi(\rho) = k\frac{\rho^2}{2R},$$

where k is the wave number and R is the radius of curvature. If the radial grid step is Δ, the phase change per step is

$$\delta\varphi(\rho) = \Delta k\frac{\rho}{R},$$

which reaches a maximum at the aperture edge $\rho = D/2$, where D is the beam diameter. Correct sampling requires $\delta\varphi < \pi$, giving

$$\Delta_{\max} < \frac{R\lambda}{D}. \tag{14.43}$$

For small R, condition (14.43) demands prohibitively fine grids, leading to very high computational cost. This difficulty can be overcome by applying a coordinate transform that replaces propagation of a spherical wave with equivalent propagation of a plane wave with initial curvature $R' = \infty$.

To illustrate, consider Fig. 14.1. An initial spherical wave of radius R_1 propagates over a distance z, acquiring a new curvature radius $R_2 = R_1 - z$. Suppose the wave then passes through a lens of focal length R_1 at a distance of $2R_1$ from the plane $z = 0$. The lens collimates the diverging beam into a plane wave, which can be propagated in Cartesian coordinates. Planes at distances z_1 and z_2 in this collimated beam correspond to images of spherical waves with radii R_1 and R_2, respectively. Thus, spherical propagation over distance z can be replaced by plane-wave propagation over a transformed distance z', followed by rescaling of amplitude and coordinates.

From lens imaging relations, the equivalent distance z' is

$$z' = \frac{R_1 z}{R_1 - z}. \tag{14.44}$$

The propagated field intensity is multiplied by

$$\frac{R_1}{R_1 - z}, \tag{14.45}$$

and the transverse coordinates x, y (and grid step Δ) are scaled by $(R_1 - z)/R_1$. For a two-dimensional field, the intensity scaling factor is $((R_1 - z)/R_1)^2$.

Thus, the algorithm is:

- Replace the spherical wave

  $$A\mathrm{e}^{\mathrm{i}(\phi(\rho)+k\rho^2/(2R))}$$

  with the plane wave

  $$A\mathrm{e}^{\mathrm{i}\phi(\rho)}.$$

- Propagate the plane wave to distance z' given by (14.44).
- Scale the resulting intensity and transverse coordinates accordingly.

The propagation can then continue in the transformed system. If a wave of curvature radius R enters a lens of focal length F, the new curvature is $RF/(R+F)$, and a new transformed system can be defined.

Coordinate transforms thus enable efficient propagation through complex lens systems: geometric optics describes the beam transforms, while diffraction propagation accounts for interference and diffraction effects. If the simulation returns to Cartesian coordinates, the quadratic phase term $k\rho^2/(2R)$ should be reintroduced, where R is the radius of curvature at the last plane.

This approach extends paraxial wave propagation methods to situations otherwise requiring extremely fine grids. However, the scalar-field paraxial approximation limits its applicability to systems with relatively small numerical apertures.

14.6 LightPipes Package

Diffraction simulation tools are available in a number of optical design environments. A free package for the simulation of coherent optical systems was developed by the author.

LightPipes is a portable collection of Unix-style tools written in C, designed to model light propagation in coherent optical devices under the scalar approximation. The toolbox is based on efficient propagation algorithms and offers extensive capabilities for beam manipulation. It enables simulations of interferometers, holographic setups, laser resonators, lasers, Fourier optics, and waveguides. The source code is highly portable and can be compiled on virtually any workstation or PC running Unix.

The package is structured as a set of filters. Each filter represents an optical element or a propagation step. In a Unix shell environment, filters can be connected

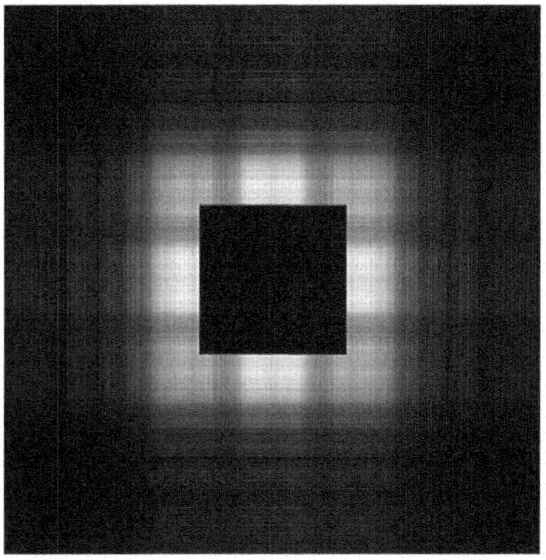

Fig. 14.2 Simulated intensity distribution of the fundamental mode at the output of an unstable resonator, obtained using a LightPipes model

in arbitrary order using system pipes to construct complete models of optical systems. The exact syntax depends on the shell used.

All filters operate on a common data structure containing a sampled two-dimensional distribution of the scalar complex amplitude. These data structures can be saved to or retrieved from disk at any stage of computation. The number of beams that can be processed simultaneously is limited only by available disk space. The package also includes utilities to generate presentation-quality PostScript density plots, export data in portable graymap (PGM) format, and produce surface wire-frames.

The original LightPipes source code in standard C and port to Matlab are available at www.okotech.com. Port to Python is available on GitHub [6].

Below is a sample script written for the original Unix version of LightPipes, modeling an unstable laser resonator. The corresponding output is illustrated in Fig. 14.2.

```
#!/bin/csh -f
# LightPipes script: model of an unstable resonator
# with N=10, M=2
# See D.B. Rench, Applied Optics 13, 2546âŁ"2561 (1974)
# Dimensions taken from the table on p. 2552
# Theoretically the initial field distribution may be random,
# but here we use a plane wave to ensure faster convergence.

./begin 20. 3e-4 1024 >field1

# Perform 20 field iterations inside the resonator
@ i=0
while( $i < 20)
# Reflection from convex mirror and propagation
./rect_ap 5.48 < field1 | ./lens -1e4 | ./forvard 1e4 > field2
# Reflection from concave mirror and propagation
./rect_ap 10.96 < field2 | ./lens 2e4 | ./forvard 1e4 | ./normal y > field1
./Strehl y < field1 >/dev/null
@ i += 1
end
```

```
# Output of the resonator screened by the output mirror
./rect_screen 5.48 < field1 | ./file_int in | \
./file_pha pha | ./file_ps res_out.ps | ./Strehl y >/dev/null

# Remove temporary files
rm field1 field2
```

14.7 Conclusion

Two complementary approaches are introduced for simulation of light propagation.

The spectral method expands the field into plane-wave components, propagates each by a phase factor, and reconstructs the field by inverse Fourier transform. Implemented with FFT algorithms, this approach can be used to model Fresnel and Fraunhofer diffraction, interference, and free-space beam propagation. For propagation in uniform isotropic medium, the FFT method offers an accurate and computationally efficient solution.

The finite-difference solution of the parabolic equation offers greater flexibility for inhomogeneous or absorbing media. Inclusion of spatially varying refractive index or absorption is straightforward, although backscattering and interface reflections are not fully accounted for. Accuracy of the method depends on step size in both transverse and longitudinal directions, with absorbing boundary layers reducing artificial boundary reflections, setting a more realistic boundary conditions.

By mapping spherical or parabolic waves into equivalent plane-wave propagation, coordinate transformations avoid prohibitively fine sampling and enable efficient modeling of complex optical systems with divergent and convergent beams.

14.8 Problems

14.1 Define complex coherent field $U(x, y) = \text{const}$, with wavelength $\lambda = 0.63\,\mu\text{m}$, in a linear grid of 1024 points with physical size of 5 mm. Define an aperture with width of 2 mm and propagate the field to the distance of 1 m using FFT method. Visualize the result.

Do the same with finite-difference method.

Do the same, but instead of free space propagation, propagate the field through a lens with focal length of 2 m, to the focus of the lens.

Solution (14.1). Matlab script realizing wave propagation using the finite difference method to solve parabolic equation numerically as described in Sect. 14.4, 14.4.1:

```
clear all;
% initial data, all sizes in meter
```

```
lambda=0.63e-6;  wave_num=2.*pi/lambda; % wavelength and k
num_points=1024;  size=5e-3;  % sampling and size of the calculation area
apert=2.0e-3; % aperture size
dz=0.01;  Z=1;  % step on Z, and total distance Z to propagate
z_foc=Z/2;  % distance to focal point in the example with converging wave
nz=Z/dz;  % number of Z steps
dx=size/(num_points-1.); % step size
nc=num_points/2+1;    % center index
zz=0;
for i = 1: num_points  % Forming the initial field array:
x=(i-nc)*dx;   % coordinate in the aperture
U(i)=complex(0);  % zero padding
if abs(x) < apert/2.
U(i) = 1;
% Uncomment for spherical wave
%U(i)=exp(complex(0.,-1.) *wave_num*(x^2)/(2*z_foc));
end
end
subplot(2,2,1) % plots of the initial fiels
plot(abs(U).*abs(U));ylim([0 1.]);
title('Initial intensity');grid on; drawnow;
subplot(2,2,3) % plots of the initial fiels
plot(angle(U));title('Initial phase');grid on; drawnow;
for i=1:nz    % The main loop  nz steps of dz each
a= ones(num_points) / dx^2; % double sweep coefficients
b= ones(num_points)/ dx^2;
c= ones(num_points)*(2./dx^2-2.*complex(0.,1.)*wave_num/dz);
f= -U*2.*complex(0.,1.)*wave_num /dz;
k1=1; mu1=0; k2=1; mu2=0;  % boundary conditions
alpha(1)=k1; beta(1)= mu1;  % first boundary
for i=1:num_points-1;  % forward sweep
alpha(i+1)=b(i)/(c(i)-alpha(i)*a(i));   % forward
beta(i+1)=(a(i)*beta(i)+f(i))/(c(i)-alpha(i)*a(i));
end
% second boundary
U(num_points)=( k2*beta(num_points)+mu2)/(1.-k2*alpha(num_points));
for i=num_points-1:-1:1 % back sweep
U(i)=alpha(i+1)*U(i+1) + beta(i+1);
end
zz=zz+dz;
% plotting the field intensity and phase
subplot(2,2,2)
plot(abs(U).*abs(U)); xlabel(['distance = ', num2str(zz)]);
title('Propagated intensity');grid on;
drawnow;
subplot(2,2,4)
plot(angle(U)); title('Propagated phase');grid on;
drawnow;
end
```

Matlab script realizing wave propagation using the FFT spectral method as described in Sect. 14.3 (Figs. 14.3 and 14.4):

```
%Propagation of a plane wave in  2D
clear all;
% initial data, all sizes in meter
lambda=0.63e-6; % wavelength
wave_num=2.*pi/lambda; % wave number k
num_points=1024; % sampling dimension
size=5e-3;  % total size of the calculation area
```

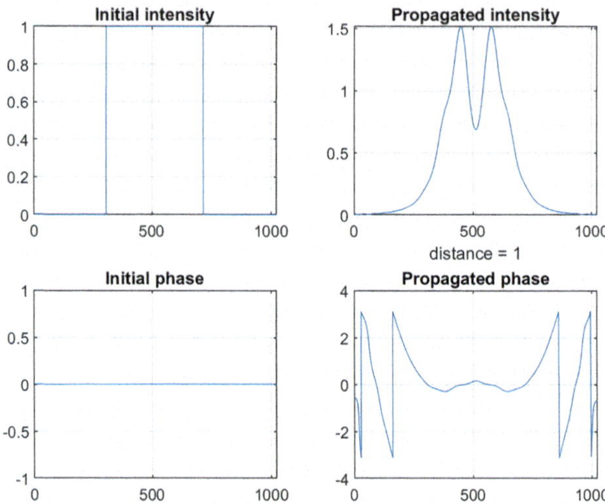

Fig. 14.3 Output of the Matlab routine realizing the finite difference method

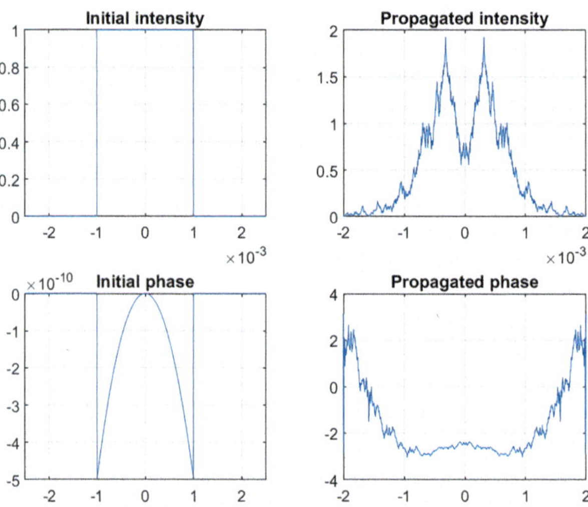

Fig. 14.4 Output of the Matlab routine realizing the FFT spectral method. The initial phase is parabolic with very low amplitude because a parabolic wave with extremely large radius of curvature (10^{10} m) was used to approximate the plane wave

```
apert=2e-3; % full aperture size
Z=1;      % distance to propagate
Zfoc=1e10; % focal length
pitch=size/(num_points-1.); % step size
% Forming the initial field array:
U = ones(num_points,1);
% array of coordinates
xxx = sampling(U,pitch);
% parabolic wavefront
U = parabolic_phase(U, pitch,lambda,Zfoc );
U = strip_aperture(U,pitch,apert/1.0);
% plotting the initial fields
subplot(2,2,1)
plot(xxx,abs(U).*abs(U)); %
title('Initial intensity');grid on
subplot(2,2,3)
plot(xxx,angle(U));
title('Initial phase');grid on

U = strip_propagate(U,pitch,lambda,Z);
subplot(2,2,2)
plot(xxx,abs(U).*abs(U));
xlim([-apert apert]);
title('Propagated intensity');grid on;
subplot(2,2,4)
plot(xxx,angle(U));
xlim([-apert apert]);
title('Propagated phase');grid on
drawnow;
function[UU] = strip_aperture(U,pitch,aperture)
num_points = length(U);
size = pitch*(num_points - 1);
nc = num_points/2+1;
UU = U;
for i = 1: num_points
x=(i-nc)*pitch;   % coordinate in the aperture
if abs(x) > aperture/2.0 UU(i) = 0; end
end
end
function[UU] = strip_propagate(U,pitch,lambda,Z)
wave_num = 2.0*pi/lambda;
num_points = length(U);
size = pitch*(num_points - 1);
nc = num_points/2+1;
Uf=fftshift(fft(fftshift(U)));  % FFT transform of the initial field
for i = 1:num_points
ic=i-nc;
serv = Z*sqrt(wave_num^2-(wave_num*ic*lambda/size)^2);
ff(i)= exp(complex(0.,1)* serv);    % Free space filter
end
UU=fftshift(ifft(fftshift(Uf .* ff))); % filter applied and inverse FFT
end
function[UU] = parabolic_phase(U,pitch,lambda,F)
wave_num = 2.0*pi/lambda;
num_points = length(U);
size = pitch*(num_points - 1);
nc = num_points/2+1;
for i = 1: num_points
x=(i-nc)*pitch;   % coordinate in the aperture
UU(i)=U(i)*exp(complex(0.,-1.)*wave_num*x^2/2.0/F);
end
end
```

References

1. J.D. Schmidt, *Numerical Simulation of Optical Wave Propagation with Examples in MATLAB* (Press Monograph, SPIE, 2010). isbn: 9780819483263
2. J.B. Breckinridge, D.G. Voelz, *Computational Fourier Optics: A MATLAB Tutorial* (SPIE Press Monograph, SPIE Press, 2011). isbn: 9780819482044
3. M.L. Calvo, V. Lakshminarayanan, *Optical Waveguides: From Theory to Applied Technologies* (Optical Science and Engineering, CRC Press, 2018). isbn: 9781420017779
4. A.A. Samarski, A.V. Gulin, *Numerical Methods* (In Russian) (Nauka, 1989)
5. G. Vdovin, *LightPipes: beam propagation toolbox for Unix, MatLab and MathCAD— Software—Products—okotech.com*. `https://www.okotech.com/lightpipes`
6. *GitHub—opticspy/lightpipes: LightPipes for Python, "Pure Python version"—github.com*. https://github.com/opticspy/lightpipes

Appendix A
Important Mathematical Formulas

Sum of geometrical progression:

$$\sum_{k=0}^{n-1} ar^k = a\frac{1-r^n}{1-r} \tag{A.1}$$

Binomial theorem:

$$(a+b)^n = \sum_{r=0}^{n} \frac{n!}{r!(n-r)!} a^{n-r} b^r. \tag{A.2}$$

Taylor series:

$$f(x) = f(a) + f'(a)\frac{x-a}{1!} + f''(a)\frac{(x-a)^2}{2!} + \cdots + f^{(n-1)}(a)\frac{(x-a)^{n-1}}{(n-1)!} + \ldots \tag{A.3}$$

Complex numbers:

$$\begin{aligned} i^2 &= -1 \\ e^{ix} &= \cos x + i \sin x \\ \cos x = \frac{e^{ix}+e^{-ix}}{2}, \quad \sin x &= \frac{e^{ix}-e^{-ix}}{2i} \\ a + ib &= \sqrt{a^2+b^2} e^{i\,\arctan(b/a)} \end{aligned} \tag{A.4}$$

G. Vdovin, *Elementary Technical Optics*, UNITEXT for Physics,
https://doi.org/10.1007/978-3-032-08626-6

Index

G. Vdovin, *Elementary Technical Optics*, UNITEXT for Physics,
https://doi.org/10.1007/978-3-032-08626-6